GCSE CHEMISTRY PRACTICAL ASSESSMENT

TEACHER'S MANUAL

with
Laboratory Technician's Notes

by Brian Higginson

HUTCHINSON

London • Melbourne • Auckland • Johannesburg

Hutchinson Education

An imprint of Century Hutchinson Ltd
62–65 Chandos Place, London WC2N 4NW

Century Hutchinson Australia Pty Ltd
PO Box 496, 16–22 Church Street, Hawthorn,
Victoria 3122, Australia

Century Hutchinson New Zealand Limited
PO Box 40–086, Glenfield, Auckland 10,
New Zealand

Century Hutchinson South Africa (Pty) Ltd
PO Box 337, Bergvlei, 2012 South Africa

First published 1988

© Brian Higginson 1988

Set in Bembo by 🅐 Tek Art Ltd, Croydon, Surrey

Printed and bound in Great Britain

British Library Cataloguing in Publication Data

Higginson, Brian
  GCSE chemistry practical assessments.
  Teacher's guide
  1. Chemistry
  I. Title
  540   QD33

ISBN 0-09-172979-3

CONTENTS

Chapter 1

Introduction

The assessment of skills is a problem which has been highlighted by the arrival of GCSE. Many teachers have evaluated their pupils' development of skills before GCSE, but now the assessment is formalised. All teachers are able to say, 'I know he/she is good at . . .'. Now we need to be able to say, 'I know he/she is good at . . . and what's more, I can prove it.'

The other problem which GCSE creates is that of comparable standards between one school and another, one class and another, and between fellow pupils. An analysis of the different examination groups reveals little difference in expectations with regard to the assessment of skills, quite rightly so because they are all based on the same National Criteria. A summary of these skills and a comparison of mark schemes is given in Chapter 3.

This scheme provides assignments of different standards, which should help teachers discriminate between less able and more able pupils within a group, and facilitate comparison of different groups within the school. It should be left to the teacher's professional judgement to decide which assignments are applicable.

There are six major areas of skills' assessment to be carried out by teachers using this scheme. These skills are combined by the different examination groups in various ways, but essentially they cover all aspects required. Teachers must group together the skill areas used in this scheme to fit the requirements of their particular examination group.

The six skill areas used in this scheme are:

1. *Following instructions*
2. *Manipulation*
3. *Making and recording observations*
4. *Drawing conclusions*
5. *Presenting data*
6. *Designing experiments*

Assessment of skills

All examination groups demand that assessment should take place as part of the normal teaching/practical process. The assignments given in this scheme are provided to satisfy that demand. They have been selected because they are part of the teaching syllabus prescribed. Also, they are relatively inexpensive to carry out. They also lend themselves to the actual learning process involved in skills' assessment.

The marking schemes for individual assignments are contained in Chapter 4 of this manual, along with advice for teachers and laboratory technicians.

Arranging the laboratory

These assignments are designed to ensure that all pupils can participate without always being assessed. It is likely that only a few pupils will be assessed at any one time, particularly for the skill areas *Manipulation* and *Following instructions*, where the teacher has to observe the pupils' actions very closely.

This has implications on the layout of the laboratory, because different groups may be working under different conditions. This requires the skills of both teacher and laboratory technician, whose job it is to prepare the laboratory for each assignment. Equipment should be set up in advance at specific 'stations' within one area of the laboratory for those pupils being assessed, unless otherwise instructed. Other pupils can collect their own equipment at the start of the assignment and work elsewhere in the laboratory. It would put unfair pressure on those pupils being assessed to have to select their own equipment (unless, of course, this skill is being assessed), when they are trying to concentrate on other skills.

The role of the laboratory technician

The advent of GCSE has meant many changes in the routine of laboratory technicians. They have no doubt been aware of the rapid change in demand on their time, and their ability to be in five places at once.

This scheme aims to assist the laboratory technician. Chapter 4 gives lists of equipment and material for each assignment, where applicable, and indicates some of the problems which may occur. Advance planning is essential if the assignments are to go smoothly, so teachers should get together with their laboratory technicians at least one week before each assignment to discuss what needs to be done. Discussions afterwards can also help to bring to light any problems which may have occurred, or suggestions for improvements in future assignments.

The laboratory technician is responsible for equipment being clean and in good working order, and the laboratory being prepared as required in advance (see above).

The skill areas

Each of the six skill areas outlined earlier should be assessed independently from the others. A poor assessment of one skill should not infringe on the assessment of any other skill. For example, if a pupil has a limited skill in *recording* data, this should not adversely affect his opportunity to show his skill at *drawing* a graph. In this case, the teacher would have to have other data available if the pupil fails to obtain his own.

A summary of skills and assignments is given in Chapter 2. Marking schemes with detailed criteria for each of the skills being assessed are given for each assignment in Chapter 4.

Following instructions

This skill is fundamental to all laboratory work; an inability to follow instructions inevitably affects performance. The instructions given in these assignments, therefore, have been carefully prepared and are backed up with diagrams. There is no reason at all why teachers should not read out instructions before the assessment starts, or during the assessment if a pupil finds difficulty reading them. Remember, the assessment is not about the pupil's ability to read English – that will be assessed by other colleagues.

If a pupil has to have instructions explained then he or she will be able to achieve only lower grades for this skill. However, once instructions are explained and understood, the pupil should then be able to go on and achieve a high grade in *Manipulation*, for example.

It is particularly important for this skill that pupils being assessed are grouped close enough together for the teacher to be able to watch all of them closely, but far enough apart to maintain examination conditions.

Manipulation

It is important that the assessment of this skill is not confused with *Following instructions*. Both can be assessed simultaneously by direct observation and very often the criteria appear to be the same. For instance, 'places the clamp' can mean that the clamp has been manipulated correctly, or that the instruction has been carried out. It could be argued that if the clamp has been manipulated correctly then an instruction has been carried out. The reverse is almost certainly not true. For example, it is possible to obey the instruction for 'pouring' with a very poor manipulative skill.

Such manipulative skills should be carried out safely, and this can be assessed only by direct observation by the teacher.

As for *Following instructions*, pupils being assessed for this skill should be close enough for the teacher to observe, but far enough apart for maintaining examination conditions.

Making and recording observations

This assessment is usually based on the pupil's written performance, but it has to be the pupil's own work – either from their own experiment or observing a teacher-demonstrated experiment.

Making and recording observations can usually be linked to the skill *Drawing conclusions*. However, teachers should always check pupils' results before asking for a conclusion to be made. Incorrect observations made during an experiment should not affect a pupil's ability to make a good conclusion.

Drawing conclusions

This skill is probably the most difficult one for pupils to achieve, so they should be given adequate practice at such assignments before they are actually observed. It can be related directly to *Making and recording observations*.

Wherever possible, conclusions should be drawn from the pupil's own work. If that work is incorrect or inadequate, the teacher should correct the pupil's work or supply him with adequate information.

These assignments are designed to point pupils in the right direction rather than helping them actually to make the desired conclusions, which would affect their skill level achieved. In other words, giving assistance usually results in a lower skill level being awarded. This means that the teacher has to check a pupil's results while the assignment is in progress, and thus check that the experiment is working. A teacher should give assistance only if a pupil is having obvious problems with arriving at a conclusion.

Presenting data

Data can be presented both in tables and in graphs. These skills are assessed by written performance, often in the same assignment (i.e. graphs drawn from tables of recordings).

The design of a table is important for conveying information clearly and concisely. A table which only the pupil can understand is not a good table, even if the information it contains is accurate. If a pupil has difficulty in devising a reasonable table of his own, then the teacher should either help or provide him with one containing the correct headings and columns. (Some of these are provided in Chapter 4). The pupil will then achieve a lower skill level for presenting a table, but will not be penalised during the rest of the assessment (e.g. in drawing graphs and making conclusions).

Some examination groups insist that this skill is assessed by using a set of collated results. It may be useful for all teachers to attempt this from time to time, as this will help him compare pupils and different class groups. This is useful when external moderation is required.

Designing an experiment

At first this seems to be the most difficult skill to assess. However, if sufficient guidance is given it should be possible for all pupils to show some skill in this area.

Although the assessment of this skill is based on written performance, it is important that all pupils are given the opportunity to carry out their designs where feasible. Teachers should encourage their pupils to use the equipment listed on the worksheet, but should not penalise them if they come up with other ideas. This means that a wide variety of items may be required; it would be impossible to list all of these in the *Laboratory technician's notes* sections of Chapter 4 and therefore it is best left for the teacher to decide.

These assignments are of two kinds. The first is the more traditional one, such as designing an experiment to show that a black powder is a catalyst. The second is of a problem-solving nature, and these assignments are related to real-life situations, such as using chromatography to find a rogue dye in a set of felt-tipped pens.

There is much pleasure for pupils being assessed in this skill: it is the one time they are asked to use their imagination – something often overlooked by science teachers in the past.

Skills: A summary

Skill area	Method of assessment	Numbers assessed
Following instructions	Teacher observation	8–10 max.
Manipulation	Teacher observation	8–10 max.
Making and recording observations	*Written performance under examination conditions	20+
Drawing conclusions	*Written performance based on own experiment, under examination conditions	20+
Presenting data	*Written performance based on own experiment, under examination conditions	20+
Designing experiments	Written performance, under examination conditions	20+

*Wherever possible, performance should be based on the students' own results from their own practical work.

Subject areas of skill assessments

Records of achievement

The skill assessments in this text are designed to be versatile. They can be used in modular science courses, and in 'broad and balanced' science courses. They should also prove useful in the assessment of skills for records of achievement.

Chapter 2

The Assignments – A Summary

ACIDS, BASES AND SALTS

1. Investigating the reactions of carbonates with acids and heat
   Skills: *Following instructions*
   Presenting data
   Drawing conclusions

2. Investigating the reactions of sodium hydroxide with metal salt solutions
   Skills: *Making and recording observations*
   Drawing conclusions

3. Finding the rules for the solubility of salts
   Skills: *Presenting data*
   Drawing conclusions

4. Preparing sulphates
   Skills: *Following instructions*
   Making and recording observations

5. Preparing chlorides
   Skills: *Following instructions*
   Making and recording observations

6. Preparing nitrates
   Skills: *Following instructions*
   Making and recording observations

7. Preparing carbonates
   Skills: *Following instructions*
   Making and recording observations

8. Investigating the heat involved in a chemical reaction
   Skills: *Following instructions*
   Presenting data

METALS AND NON-METALS

9. Investigating the burning of metals and non–metals
   Skills: *Making and recording observations*
   Manipulation

10. Reacting metals with acids
    Skills: *Following instructions*
    Making and recording observations
    Manipulation

11. Obtaining a metal from its oxide
    Skills: *Manipulation*
    Following instructions

12. Investigating the rate of reaction between a metal and an acid
    Skills: *Making and recording observations*
    Presenting data
    Drawing conclusions

13. Investigating the displacement of metals
    Skills: *Making and recording observations*
    Drawing conclusions

14. Designing an experiment to place a metal in the reactivity series
    Skills: *Designing an experiment*

15. Separating materials into conductors and non-conductors
    Skills: *Designing an experiment*

16. Investigating a mineral
    Skills: *Following instructions*
    Manipulation
    Making and recording observations

17. Making electricity from chemicals
    Skills: *Making and recording observations*
    Manipulation

SOLIDS, LIQUIDS AND GASES

18. Finding the freezing point of water
    Skills: *Following instructions*
    Presenting data in graphs and tables

19. Determining the boiling point of a liquid
    Skills: *Following instructions*
    Manipulation

20. Using melting points to find impurities
    Skills: *Presenting data*

21. Finding the melting point of a solid
    Skills: *Following instructions*
    Presenting data
    Manipulation

SEPARATIONS, PURIFICATIONS AND POLLUTION

22. Investigating fermentation
    Skill: *Designing an experiment*

23. Using chromatography to identify an impurity
    Skill: *Designing an experiment*

24. Investigating river pollution
    Skill: *Designing an experiment*

25. Investigating the pollution near a sulphuric acid plant
    Skill: *Designing an experiment*

REACTION RATES

26. Investigating the rate of a reaction
    Skill: *Designing an experiment*

27. Investigating the rate of a reaction by loss of mass
    Skill: *Presenting data*

PREPARATION AND PROPERTIES OF GASES

28. Preparing oxygen to investigate a catalyst
    Skill: *Designing an experiment*

29. Investigating gases
    Skill: *Drawing conclusions*

30. Preparing carbon dioxide gas
    Skills: *Following instructions*
    Manipulation
    Drawing conclusions

31. Investigating burning
    Skills: *Following instructions*
    Manipulation
    Making and recording observations

WATER

32. Investigating the hardness of water
    Skill: *Designing an experiment*

33. Testing the hardness of various waters
    Skills: *Making and recording observations*
    Drawing conclusions

34. Testing water for the presence of sulphates and chlorides
    Skills: *Following instructions*
    Making and recording observations

35. Investigating rusting
    Skill: *Designing an experiment*

36. Investigating the causes and prevention of rusting
    Skills: *Following instructions*
    Manipulation
    Making and recording observations
    Drawing conclusions

37. Plotting solubility curves
    Skill: *Presenting data*

ELECTROCHEMISTRY

38. Investigating the effects of electrolysing purified sea water
    Skills: *Following instructions*
    Manipulation

39. Investigating the electrolysis of water
    Skills: *Presenting data*
    Drawing conclusions

40. The electrolysis of copper(II) chloride solution
    Skills: *Following instructions*
    Manipulation

FUELS

41. Distilling petroleum
    Skills: *Presenting data*
    Drawing conclusions

42. Investigating the cost, efficiency and cleanness of various fuels
    Skill: *Designing an experiment*

43. Distilling coal
    Skills: *Following instructions*
    Manipulation

MISCELLANEOUS AND DESIGN

This section is mainly for pupils who have problems with their assessments. The experiments have no theme apart from the fact that they lend themselves to assessment at the end of a course. They can still be used as part of the general assessment.

44. Investigating diffusion of gases
    Skill: *Presenting data*

45. Designing an experiment
    Skill: *Designing an experiment*

46. An industrial study: air liquefaction
    Skill: *Drawing conclusions*

47. Finding the number of molecules of water in a salt
    Skill: *Following instructions*

48. Analysing an unknown chemical
    Skills: *Following instructions*
    Manipulation
    Making and recording observations
    Drawing conclusions

Analysis of skills assessed

Assignment	Following instructions	Manipulation	Making and recording observations	Drawing conclusions	Presenting data	Designing an experiment	
1	■			■			1. Carbonates with acids and heat
2			■		■		2. Sodium hydroxide with metal salts
3			■	■			3. Solubility of salts
4	■		■				4. Preparing sulphates
5	■						5. Preparing chlorides
6	■		■				6. Preparing nitrates
7	■						7. Preparing carbonates
8	■				■		8. Heat in chemical reaction
9			■				9. Burning metals and non-metals
10	■	■	■				10. Reacting metals with acids
11	■			■			11. Obtaining metal from oxide
12			■	■			12. Rate of reaction
13			■		■		13. Displacement of metals
14						■	14. Reactivity series
15			■			■	15. Conductors and non-conductors
16	■						16. Investigating a mineral
17		■	■				17. Electricity from chemicals
18	■	■		■			18. Freezing point of water
19	■						19. Boiling point of liquid
20	■				■		20. Melting points and impurities
21	■	■			■		21. Melting point of solid
22						■	22. Fermentation
23						■	23. Chromatography
24						■	24. River pollution
25						■	25. Industrial pollution
26						■	26. Rate of reaction
27					■	■	27. Rate of reaction by loss of mass
28						■	28. Oxygen and catalyst
29				■			29. Investigating gases
30	■	■		■			30. Preparing carbon dioxide
31	■	■					31. Investigating burning
32			■			■	32. Hardness of water
33			■			■	33. Hardness of various waters
34	■	■	■				34. Sulphates and chlorides in water
35			■				35. Investigating rusting
36	■	■					36. Causes and prevention of rusting
37					■		37. Plotting solubility curves
38	■	■					38. Electrolysing sea water
39				■			39. Electrolysis of water
40	■			■			40. Electrolysis of copper chloride
41				■			41. Distilling petroleum
42						■	42. Investigating various fuels
43	■						43. Distilling coal
44					■		44. Diffusion of gases
45						■	45. Designing an experiment
46				■			46. Air liquifaction
47	■						47. Molecules of water in a salt
48	■						48. Analysing an unknown
Total number of times assessed	21	13	17	13	12	11	

Chapter 3

Teacher Marking Schemes

The marking schemes used for these assignments are based on points awarded to pupils who satisfy specific criteria. These criteria are then interpreted accordingly. General criteria for each of the six skills are given in this chapter; more specific criteria for individual assignments are given in Chapter 4.

Any marking scheme, whatever its nature, is always somewhat arbitrary. But hopefully these marking schemes (which are a composition of all the criteria given by each examination group) will guide the teacher to making reasonably valid judgements on the skill levels achieved by their pupils.

Skill levels

The examination boards fall into two main categories. The Northern Examining Association (**NEA**), London and East Anglian Group (**LEAG**) and Northern Ireland Schools Examinations Council (**NISEC**) all use a 1–5 point scale, with 5 being the highest level of skill. The Southern Examining Group (**SEG**), Midland Examining Group (**MEG**) and the Welsh Joint Education Committee (**WJEC**) use a 1–3 point scale, with 3 being the highest skill level.

The only exception to these categories is the WJEC which uses a somewhat arbitrary 1–2 point 'take-it-or-leave-it' scale for the skill area *Following instructions* (Skill 1.1 Routine operations and safety). This is not meant as a criticism of the WJEC, but it does not give much scope for positive achievement.

This apart, the examination groups have major areas of overlap. It is thus possible to present the following broad areas of criteria for awarding skill levels.

Following instructions

The pupil is able to:	NEA/LEAG/NISEC	SEG/MEG	WJEC
follow instructions without further guidance, conducting practical work in a logical sequence and safely	5	3	2
follow instructions with the absolute minimum of further guidance and is then able to proceed as above	4	2	2
follow instructions only after further guidance has been given, and occasional intervention	3	2	1
follow instructions with considerable assistance unless the task is fairly simple	2	1	1
follow instructions for only a simple task and still require assistance	1	1	1

Manipulation

The pupil is able to:	NEA/LEAG/NISEC	SEG/MEG/WJEC
handle all apparatus and materials in a safe and proper manner, without risk to self, fellow workers and equipment	5	3
handle most apparatus without usually requiring assistance, in a safe manner	4	2
handle apparatus in a safe manner but sometimes requires assistance to avoid further problems	3	
handle fairly simple apparatus without problem, but safety now becomes a problem	2	1
use simple apparatus provided safety assistance is given	1	

Making and recording observations

The pupil is able to:	NEA/LEAG/NISEC	SEG/MEG/WJEC
make correct observations and record them from fairly complex situations without assistance	5	3
make correct observations and record them with occasional errors unless given minimal assistance in complex experiments	4	2
make and record observations, with assistance in more complex experiments only	3	
make and record observations in simple experiments	2	1
make and record observations in simple experiments, but with errors unless given assistance	1	

Drawing conclusions

The pupil is able to:	NEA/LEAG/NISEC	SEG/MEG/WJEC
make sensible and logical conclusions based on unguided evidence or any assistance, in complex experiments	5	3
make sensible and logical conclusions but sometimes requires minimal assistance to do so in complex experiments	4	2
make conclusions in most experiments provided assistance is given	3	
make conclusions in simple experiments but may require minimal guidance	2	1
make conclusions in simple experiments provided guidance is given	1	

Presenting data

The pupil is able to:	NEA/LEAG/NISEC	SEG/MEG/WJEC
present data in a logical and sensible manner without any assitance, and complete graphs with labelled axes, good shape to curve and use of scale	5	3
present data in a logical and sensible manner from complex experiments provided some minimal assitance is given	4	2
complete, with assistance, most tasks for plotting graphs and tabulating	3	
produce, from simple experiments, a graph or table which is useful and contains most information	2	1
produce a graph or table from simple experiments only with assistance	1	

Designing experiments

The pupil is able to:	NEA/LEAG/NISEC	MEG/WJEC (SEG)*
design an experiment to test a theory in a sensible logical fashion, given only the required background information. Can foresee errors and problems which may occur and suggest methods to overcome them	5	3
design an experiment to test a theory in a sensible, logical way, given the background information and minimum assistance. Can foresee errors and problems and suggest ways to overcome them	4	2
design an experiment to test a theory in a sensible logical way, but requires assistance	3	2
design a simple experiment to test a simple theory	2	1
design an experiment to test a simple given theory but only with assistance	1	1

*External assessment of this skill is not required by the SEG.

Chapter 4

The Assignments – In Detail

Assignment 1
Investigating the reactions of carbonates with acids and heat

Teacher's notes

Pupils are expected to be familiar with the test for carbon dioxide gas and to know that this is usually acceptable evidence for the existence of a carbonate. As this assignment is not about knowledge or skill manipulation as such, you could demonstrate the test for carbon dioxide before the pupils start the assessment.

The assignment is designed as a full group activity. The skills *Presenting data* and *Drawing conclusions* can be measured by the pupils' written performance at the end of the assessment. The skill *Following instructions* can be measured only under supervision by the teacher. A check list is provided to assist you in carrying out this part of the assessment (see page 21).

In the laboratory technician's notes (see page 20) it is suggested that six carbonates are provided. This means that it should be possible to follow the work of at least six candidates very closely, or more if necessary.

Marking schemes

Skill: Following instructions

Criteria:	*Mark*
Wears goggles throughout experiment	1
Uses holders and correct response (Stage 2)	1
Uses 5 ml of limewater	1
Heats carbonate as shown	1
Takes fresh sample at Stage 7	1
Adds acid as instructed	1
Maximum mark	6

Mark Obtained	Skill Level		
	NEA/LEAG/NISEC	MEG/SEG	WJEC
5 and 6	5	3	2
4	4		
3	3	2	1
2	2		
1	1	1	

Skill: Presenting data

Criteria: *Mark*

Single table 1
Uses columns 1
Columns are correctly titled 1
Table is well-presented and tidy 1
Table contains sensible sets of entries.............. 1
Table is logically presented........................... 1
Maximum mark 6

Mark Obtained	Skill Level	
	NEA/LEAG/NISEC	MEG/SEG/WJEC
5 and 6	5	3
4	4	
3	3	2
2	2	
1	1	1

Skill: Drawing conclusions

Criteria: *Mark*

All carbonates give off carbon dioxide gas with acids 3
As above but with some minimal guidance 2
As above but with a great deal of assistance 1

All carbonates, except _____, give off carbon dioxide
   gas when they are heated... 3
As above but with assistance 2
As above but with much assistance...................................... 1
Maximum mark 6

Mark Obtained	Skill Level	
	NEA/LEAG/NISEC	MEG/SEG/WJEC
5 and 6	5	3
4	4	
3	3	2
2	2	
1	1	1

Laboratory technician's notes

The materials to be used in this experiment should be available at the start of the lesson. The teacher will also find your presence in the laboratory helpful during the assignment. Each pupil being assessed will require:

> safety glasses/goggles
> clean test tubes and rack
> small bottle of hydrochloric acid (approx. 1M)
> access to a supply of limewater
> test tube holders
> spatula
> bunsen burner
> the following carbonates (or similar): sodium-, calcium-, zinc-, magnesium-, copper-
> and lead-carbonate. These should be labelled clearly and placed in a part of the
> laboratory near the pupils being assessed.

It would be very useful if five or six positions are set up for individual candidates being assessed so that the teacher can watch them. Other pupils may be able to fend for themselves for their own equipment.

X1

Pupil's Name	Wears goggles through-out experiment	Uses holders, correct response	Uses 5 ml limewater	Heats carbonate as shown	Takes fresh sample at Stage 7	Adds acid as instructed	Total Mark	Skill Level
	1	1	1	1	1	1		

Assignment 2
Investigating the reactions of sodium hydroxide with metal salt solutions

Teacher's notes

Pupils will require some prior knowledge for this assignment. They should be able to recognise a precipitate and the word. If not, you should explain this before starting the assignment, or demonstrate the formation of a precipitate.

Pupils should be provided with six salt solutions: a potassium salt, a calcium salt, a magnesium salt and three other coloured salt solutions; all clearly labelled.

This experiment requires limited apparatus and provides written answers. It should be possible to perform this assignment with a large group of pupils under examination conditions.

You should point out to the pupils that sodium hydroxide is highly corrosive to skin and clothing.

Marking schemes

Skill: Making and recording observations

Criteria: *Mark*

There are six observations to be made (one for each metal salt) on the first addition of sodium hydroxide ....1 mark each

Maximum mark..........................6

Mark Obtained	Skill Level	
	NEA/LEAG/NISEC	MEG/SEG/WJEC
5 and 6	5	3
4	4	3
3	3	2
2	2	2
1	1	1

Skill: Drawing conclusions

Criteria   *Mark*

Most salts give precipitates, except sodium, etc...........3
As above with minimum assistance...........................2
As above but with great assistance............................1

Some metals, except_____ give
   coloured precipitates with sodium hydroxide 3
As above but with some guidance 2
As above but with much guidance............................1

Maximum mark................6

Mark Obtained	Skill Level	
	NEA/LEAG/NISEC	MEG/SEG/WJEC
5 and 6	5	3
4	4	
3	3	2
2	2	
1	1	1

Laboratory technician's notes

Each pupil being assessed will require the following apparatus:

safety glasses/goggles
clean test tubes and test tube rack
small bottle (beaker) of sodium hydroxide solution (approx. 1M)
a supply of six named metal salt solutions: e.g. sodium chloride, potassium nitrate, magnesium chloride, copper sulphate, an iron(II) and an iron(III) salt. They should all be made into solutions of about one mole per litre.

It would be useful if the number of required positions for assessment are set up before the lesson starts. The teacher will tell you how many pupils are being assessed. If this is a large group, pupils may have to share test tube racks, and your assistance will be required during the assessment.

Assignment 3
Finding the rules for the solubility of salts

Teacher's notes

This assignment has been designed so that a large group of candidates can have their skills in *Presenting data* and *Drawing conclusions* measured at one time. The equipment required is simple and inexpensive. There is no reason why a full class should not attempt this assignment, provided examination conditions can be maintained.

Check that pupils have made a reasonable attempt at producing their own tables before starting their assignment. If they are unable to do so, then use the table shown on page 26 to photocopy and hand out. As pupils make their observations, check that they are doing so correctly. Pupils should not be penalised for making incorrect observations as this assignment does not measure this particular skill. It would therefore be useful if you provided a set of collective results at the end of the experiment so that each pupil has the same information to form their conclusion.

Marking schemes

Skill: Presenting data

Criteria:	Mark
Any attempt at producing a tidy table	2
As above, but needs assistance	1
Table is made from columns	1
Columns are correctly titled	1
Recordings in table should be logical as in example shown later, i.e. all metals grouped	1
Each group of recordings is lined off to keep separate	1
Maximum mark	6

Mark Obtained	Skill Level	
	NEA/LEAG/NISEC	MEG/SEG/WJEC
6	5	3
5	4	3
3 and 4	3	2
2	2	1
1	1	1

Skill: Drawing conclusions

Criteria	Mark
All nitrates are soluble	1
All chlorides are soluble except _____	2
All sulphates are soluble except _____	2
All carbonates are insoluble except _____	2
Maximum mark	7

Mark Obtained	Skill Level	
	NEA/LEAG/NISEC	MEG/SEG/WJEC
7	5	3
5 and 6	4	3
4	3	2
3	2	2
1 and 2	1	1

Laboratory technician's notes

Each pupil being assessed will require the following apparatus:

large number of named metal salts★: The carbonates, chlorides, nitrates and sulphates of sodium, calcium, magnesium, zinc, lead and copper should be available in large amounts.

test tubes and test tube rack

rubber bungs

★Small plastic petri dishes are useful containers for salts.

Assignment 3

Name..Group......................Date........................

Table of results: a check list for solubilities

Name of metal	Nitrate	Sulphate	Chloride	Carbonate

Assignment 4
Preparing sulphates

Teacher's notes

This assignment can be done by the whole group for assessing the skill *Making observations*, provided each pupil can make his own set of observations independently. For the skill *Following instructions* only seven or eight pupils can realistically be assessed at any one time. However, each pupil has to repeat the instructions four times so it may be possible to increase the number of pupils being assessed (i.e. one group is assessed for metals, another for metal oxide, another for metal carbonate, etc.).

It is suggested that coloured metal salts should be made if possible, because these usually form coloured precipitates which are easy to see. The starting materials should be in powdered form, e.g. metal powder (iron), metal oxide (copper), metal carbonate (copper), metal hydroxide (copper).

A check list for *Following instructions* is provided on page 29.

Marking schemes

Skill: Following instructions

Criteria:	*Mark*
Wears safety glasses	2
Wears safety glasses after prompting	1
Step 3 correct	1
Step 4 correct	1
Step 9 correct (decanting)	1
Maximum mark	5

Mark Obtained	Skill Level		
	NEA/LEAG/NISEC	MEG/SEG	WJEC
5	5	3*	2*
4	4	3*	2*
3	3	2	2*
2	2	2	1
1	1	1	1

*Must include 2 marks for wearing safety glasses.

Skill: Making and recording observations

Criteria:	*Mark*
There are 12 observations to be made to complete the table. For any reasonable attempt at an observation	1 mark each
Maximum mark	12

Mark Obtained	Skill Level	
	NEA/LEAG/NISEC	MEG/SEG/WJEC
11 and 12	5	3
9 and 10	4	
6, 7 and 8	3	2
3, 4 and 5	2	
1 and 2	1	1

Laboratory technician's notes

Each pupil carrying out this assignment will need the following:

eight test tubes
test tube rack
small bottle (beaker) of sulphuric acid (2M) clearly labelled
powdered forms of a metal, metal oxide, metal carbonate, metal hydroxide as indicated
    by the teacher
a suitable storage place for retaining test tubes

The teacher is likely to need you in the laboratory for this assignment to assist with problems which may occur. It would be helpful if the apparatus is ready and in position at the start of the session for the pupils carrying out the assessment. Other pupils may also be doing the assignment but not for assessment purposes. These pupils will still require apparatus but they may be able to collect it themselves.

Skills check list – *Following instructions*

Pupil's Name	Wears safety glasses	Wears safety glasses after prompting	Step 3 is correct	Step 4 is correct	Step 9 is correct (decanting)	Total Mark	Skill Level
	2	1	1	1	1		

Assignment 5
Preparing chlorides

Teacher's notes

This assignment can be done by the whole group for assessing the skill *Making observations*, provided each pupil can make his own set of observations independently. For the skill *Following instructions* only seven or eight pupils can realistically be assessed. However, as in Assignment 4, each pupil has to repeat the instructions four times so it may be possible to increase the number of pupils being assessed (i.e. different pupils can be assessed for different starting materials).

It is suggested that coloured metal salts should be made if possible, because these usually form coloured precipitates which are easy to see. The starting materials should be in powdered form, e.g. metal powder (iron), metal oxide (copper), metal carbonate (copper), metal hydroxide (copper).

A check list for *Following instructions* is provided on page 29.

Marking schemes

Skill: Following instructions

Criteria:	Mark
Wears safety glasses	2
Wears safety glasses after prompting	1
Step 3 correct	1
Step 4 correct	1
Step 9 correct (decanting)	1
Maximum mark	5

Mark Obtained	Skill Level		
	NEA/LEAG/NISEC	MEG/SEG	WJEC
5	5	3*	2*
4	4		
3	3	2	
2	2		1
1	1	1	

*Must include 2 marks for wearing safety glasses.

Skill: Making and recording observations

Criteria:	Mark
There are 12 observations to be made to complete the table. For any reasonable attempt at an observation	1 mark each
Maximum mark	12

Mark Obtained	Skill Level	
	NEA/LEAG/NISEC	MEG/SEG/WJEC
11 and 12	5	3
9 and 10	4	
6, 7 and 8	3	2
3, 4 and 5	2	
1 and 2	1	1

Laboratory technician's notes

Each pupil carrying out this assignment will need the following:

> eight test tubes
> test tube rack
> small bottle (beaker) of sulphuric acid (2M) clearly labelled
> powdered forms of a metal, metal oxide, metal carbonate, metal hydroxide as indicated
> by the teacher
> a suitable storage place for retaining test tubes

The teacher is likely to need you in the laboratory for this assignment to assist with problems which may occur. It would be helpful if the apparatus is ready and in position at the start of the session for the pupils carrying out the assessment. Other pupils may also be doing the assignment but not for assessment purposes. These pupils will still require apparatus but they may be able to collect it themselves.

Assignment 6
Preparing nitrates

Teacher's notes

This assignment can be done by the whole group for assessing the skill *Making observations*, provided each pupil can make his own set of observations independently. For the skill *Following instructions* only seven or eight pupils can realistically be assessed. However, as in Assignment 4, each pupil has to repeat the instructions four times so it may be possible to increase the number of pupils being assessed (i.e. different pupils can be assessed for different starting materials).

It is suggested that coloured metal salts should be made if possible, because these usually form coloured precipitates which are easy to see. The starting materials should be in powdered form, e.g. metal powder (iron), metal oxide (copper), metal carbonate (copper), metal hydroxide (copper).

A check list for *Following instructions* is provided on page 29.

Marking schemes

Skill: Following instructions

Criteria:	Mark
Wears safety glasses	2
Wears safety glasses after prompting	1
Step 3 correct	1
Step 4 correct	1
Step 9 correct (decanting)	1
Maximum mark	5

Mark Obtained	Skill Level		
	NEA/LEAG/NISEC	**MEG/SEG**	**WJEC**
5	5	3*	2*
4	4	3*	2*
3	3	2	2*
2	2	2	1
1	1	1	1

*Must include 2 marks for wearing safety glasses.

Skill: Making and recording observations

Criteria:	Mark
There are 12 observations to be made to complete the table. For any reasonable attempt at an observation	1 mark each
Maximum mark	12

Mark Obtained	Skill Level	
	NEA/LEAG/NISEC	MEG/SEG/WJEC
11 and 12	5	3
9 and 10	4	
6, 7 and 8	3	2
3, 4 and 5	2	
1 and 2	1	1

Laboratory technician's notes

Each pupil carrying out this assignment will need the following:

    eight test tubes
    test tube rack
    small bottle (beaker) of nitric acid (about 2M) – take care
    powdered forms of a metal, metal oxide, metal carbonate, metal hydroxide as indicated
        by the teacher
    a suitable storage place for retaining test tubes

The teacher is likely to need you in the laboratory for this assignment to assist with problems which may occur. It would be helpful if the apparatus is ready and in position at the start of the session for the pupils carrying out the assessment. Other pupils may also be doing the assignment but not for assessment purposes. These pupils will still require apparatus but they may be able to collect it themselves.

Assignment 7
Preparing carbonates

Teacher's notes

It would be useful if pupils carry out Assignment 3 before this assignment is considered, otherwise pupils should have some knowledge and understanding of the rules of solubility. The skill *Making and recording observations* is assessed by written performance. This leaves you time to assess pupils' ability at *Following instructions*. A check list is provided for this skill on page 36.

A large group of candidates can do this assignment if equipment allows and examination conditions can be maintained.

It is suggested that six metal salts are provided, e.g. calcium chloride, magnesium chloride, copper sulphate, lead nitrate, iron(II) chloride, zinc nitrate solutions, or similar. The only problem with this experiment is that it uses a large amount of filter paper!

Marking schemes

Skill: Following instructions

Criteria:	Mark
Step 2 correct	1
Step 3 correct	1
Step 4 correct	1
Step 5 correct . . . apparatus	1
. . . filter paper	1
Step 6 correct	1
Maximum mark	6

Mark Obtained	Skill Level		
	NEA/LEAG/NISEC	**MEG/SEG**	**WJEC**
5 and 6	5	3	2
4	4		
3	3	2	1
2	2		
1	1	1	

Skill: Making observations

Criteria:	Mark
There are 12 observations to be made	1 mark each
Maximum Mark	12

Mark Obtained	Skill Level	
	NEA/LEAG/NISEC	MEG/SEG/WJEC
11 and 12	5	3
9 and 10	4	
6, 7 and 8	3	2
3, 4 and 5	2	
1 and 2	1	1

Laboratory technician's notes

Each pupil taking this assignment will need the following:

    some named metal salt solutions
    sodium carbonate solution
    test tubes
    test tube rack
    filter funnel
    filter papers

It will be useful if you are available in the laboratory during skills testing.

Skills check list – *Following instructions*

Pupil's Name	Step 2 correct 1	Step 3 correct 1	Step 4 correct 1	Step 5 correct ... apparatus 1	Step 5 correct ... filter paper 1	Step 6 correct 1	Total Mark	Skill Level

Assignment 8
Investigating the heat involved in a chemical reaction

Teacher's notes

This assignment can be performed by a large group of pupils, but it will be possible to assess only a small number of them for the skill *Following instructions*. The other skill being assessed is the pupil's ability to present graphs and tables. It is preferable for pupils to obtain their own results and plot their own graphs. However, if their results are not valid then you should collate the class results and ask the pupils to plot graphs from this collected data. Remember, you should not penalise a pupil in one skill for failing in another, and examination boards expect collated data to be used when necessary.

It is essential that pupils are made aware of the dangers of the materials they are using in this assignment. Part of the assessment is their ability to act in a safe manner.

Pupils should work on their own if possible, but many of the group could work in pairs as their assessment can be based on collated results and not necessarily their own.

A skills check list for *Following instructions* is provided on page 39.

Marking schemes

Skill: Following instructions

Criteria:	Mark
Step 2 correct	1
Step 3 correct	1
Step 4 correct	1
Step 5 correct	1
Step 6 correct	1
Wears safety glasses	1
Maximum mark	6

Mark Obtained	Skill Level		
	NEA/LEAG/NISEC	MEG/SEG	WJEC
5 and 6	5	3	2
4	4	2	
3	3		1
2	2	1	
1	1		

Skill: Presenting data

Criteria:	*Mark*

Axes correctly labelled (both) .... 2 (1 each)
Good use of scale on each axis 2
Most points placed correctly 1
Graph is a reasonable shape 1

Maximum mark 6

Mark Obtained	Skill Level	
	NEA/LEAG/NISEC	MEG/SEG/WJEC
5 and 6	5	3
4	4	3
3	3	2
2	2	2
1	1	1

Laboratory technician's notes

Each pupil will need the following apparatus and materials:

thermometer (0 to 110°C)
plastic cup insulated with any suitable
    material (e.g. polystyrene balls,
    cotton wool, paper, etc.) to provide
    a jacket around the cup, which
    should then be placed in a 250 ml
    beaker as shown in the diagram:
bottle of 2M sulphur acid, clearly
    labelled
bottle of 2M sodium hydroxide, clearly
    labelled
two 100 ml measuring cylinders (if
    possible)
safety glasses/goggles

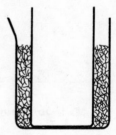

As there is acid and alkali about, you should be at hand during the assessment to help clean up any spillages.

Skills check list – *Following instructions*

Pupil's Name	Step 2 correct	Step 3 correct	Step 4 correct	Step 5 correct	Step 6 correct	Wears safety glasses	Total Mark	Skill Level
	1	1	1	1	1	1		

Assignment 9
Investigating the burning of metals and non-metals

Teacher's notes

Both skills in this assignment can be assessed only for a small number of pupils, unless examination conditions can be provided as well as the apparatus. The optimum number is 10 pupils.

Pupils should be aware of the acidic and basic nature of the indicator being used, (i.e. the colour change in response to acidic and basic materials being tested).

Drawing conclusions is not assessed as a skill for this assignment, although it would be possible to, if desired. The skill *Making and recording observations* can be assessed from the table of data presented. The skill *Manipulation* must be assessed by direct observation of each pupil. The experiment is repeated five times and it should be possible to supervise simultaneously pupils working nearby. A skills check list for *Manipulation* is provided on page 42.

Five metals/non-metals should be provided, e.g. calcium, magnesium, aluminium, carbon and sulphur.

Marking schemes

Skill: Making and recording observations

Criteria: *Mark*

There are five observations to be made from the 'What colour changes did you see' column .... 1 each

                            Maximum mark 5

Mark Obtained	Skill Level	
	NEA/LEAG/NISEC	MEG/SEG/WJEC
5	5	3
4	4	3
3	3	2
2	2	2
1	1	1

Skill: Manipulation

Criteria: Mark

Step 4 correct and carried out carefully, wearing safety glasses.............2
Step 5 correct...1
Step 6 correct, shakes carefully...1
Step 7 adds indicator correctly..1
Step 3 material must not touch bottom of gas jar 1
                                                     Maximum mark...........................6

Mark Obtained	Skill Level	
	NEA/LEAG/NISEC	MEG/SEG/WJEC
5 and 6	5	3
4	4	
3	3	2
2	2	
1	1	1

Laboratory technician's notes

Each pupil will need the following:

    metals: calcium, magnesium, aluminium or iron filings
    non-metals: carbon, sulphur
    safety glasses/goggles
    gas jar and cover slip
    an indicator
    deflagrating spoon
    bunsen burner

The apparatus and materials should be in position and ready for the pupils at the start of the assignment.

b ✗

Skills check list – *Manipulation*

Pupil's Name	Step 4 correct, wears safety glasses	Step 5 correst	Step 6 correct, shakes carefully	Step 7, adds indicator correctly	Step 3, material off bottom of gas jar	Total Mark	Skill Level
	2	1	1	1	1		

Assignment 10
Reacting metals with acids

Teacher's notes

This experiment should be demonstrated to the pupils before they start. They should be aware of the hydrogen test and this part of the assignment should be made very clear to all pupils.

The assignment can be carried out by the whole group for assessing the skill *Making and recording observations*. You must also observe the other skills if they are selected for assessment. The whole group can be split into two with each half doing one of the acids at a time. Examination conditions must prevail over the whole group if everyone does the experiment as an assessment. Of course it is possible for part of the class to carry out this experiment as a normal practical assignment, with a very small group being assessed.

The skill *Making and recording observations* can be assessed from the pupils' written response. However the skills *Manipulation* and *Following instructions* have to be assessed on the spot. They are to some degree related to each other, although it is possible to 'follow instructions' but have poor 'manipulation'. However, this assignment lends itself to the assessment of both skills. Suitable check lists are provided on pages 46 and 47.

Powdered zinc, iron and aluminium should be used with acid strength at 2M.

Marking schemes

Skill: Making and recording observations

Criteria   Marks

There are six observations to be made .... 1 each

Maximum mark 6

Mark Obtained	Skill Level	
	NEA/LEAG/NISEC	MEG/SEG/WJEC
5 and 6	5	3
4	4	2
3	3	2
2	2	1
1	1	1

Skill: Manipulation

Criteria	Mark
Step 3, is able to place materials in test tube without spillage	1
Step 4, adds acid and has test tube ready, no spillage	2
Step 5, holds test tube steadily	1
Step 6, holds test tube and passes safely through bunsen flame	1
Maximum mark	5

Mark Obtained	Skill Level	
	NEA/LEAG/NISEC	MEG/SEG/WJEC
5	5	3
4	4	3
3	3	2
2	2	1
1	1	1

Skill: Following instructions

Criteria:	Mark
Wears safety glasses as indicated	1
Adds metal to test tube	1
Adds acid carefully	1
Holds second test tube ready	1
Waits two minutes for reaction to complete	1
Passes mouth of test tube through bunsen flame	1
Maximum mark	6

Mark Obtained	Skill Level		
	NEA/LEAG/NISEC	MEG/SEG	WJEC
5 and 6	5	3	2
4	4	2	2
3	3	2	2
2	2	1	1
1	1	1	1

Laboratory technician's notes

Pupils attempting this assignment require the following items (if possible they should have their own samples of metal and bottles of acid. The metals need only be in small amounts – four or five small spatulas full):

metals – powdered zinc, iron and aluminium
2M sulphuric acid (small labelled reagent bottle, with teat pipette)
2M hydrochloric acid (small labelled reagent bottle, with teat pipette)
test tubes in a rack
safety glasses/goggles
spatula
bunsen burner and mat

This apparatus should be set out in the laboratory before the assessment starts. All bottles and other chemicals should be carefully labelled. Also, you should be present in the laboratory during this assessment because the teacher will be unable to leave the room to ask for assistance.

Skills check list – *Manipulation*

Pupil's Name	Step 3, is able to place materials in test tube without spillage	Step 4, adds acid and has test tube ready, no spillage	Step 5, holds test tube steadily	Step 6, holds test tube and passes safely through bunsen flame	Total Mark	Skill Level
	1	2	1	1		

Skills check list – *Following instructions*

Pupil's Name	Wears safety glasses as indicated	Adds metal to test tube	Adds acid carefully	Holds second test tube ready	Waits 2 minutes for reaction to complete	Passes mouth of test tube through bunsen flame	Total Mark	Skill Level
	1	1	1	1	1	1		

Assignment 11
Obtaining a metal from its oxide

Teacher's notes

This assignment requires you to make observations of the pupils' ability to manipulate apparatus as well as to follow written instructions. These observations must be done at the same time. Both skills require careful attention to safety. Check lists are provided for you to record the pupils' skills on pages 50 and 51.

You may wish to use this assignment for practising the calculation of formulae, but this is not provided for on the assignment sheet. Pupils should not be penalised for making mistakes in their weighings as this is clearly not part of the assessment. It belongs in another skill area.

Copper oxide is a useful material to use in this experiment as it provides a quick result. However, copper carbonate could be used as an alternative. If the apparatus requirements do not suit the resources of the laboratory, then feel free to change the diagrams within the text.

Marking schemes

Skill: Manipulation

Criteria:	Mark
Raises clamp correctly by unscrewing	1
Refastens screw correctly	1
Clamp is in safe position, teeth supporting	1
Places test tube in clamp to correct position	1
Tightens up clamp correctly	1
Puts copper oxide into test tube without spillage	1
Weighs tube without spillage	1
Correctly replaces tube with hole facing upwards	1
Sets gas correctly to a small and safe flame	1
Maximum mark	9

Mark Obtained	Skill Level	
	NEA/LEAG/NISEC	MEG/SEG/WJEC
8 and 9	5	3
7	4	3
5 and 6	3	2
3 and 4	2	1
1 and 2	1	1

Skill: Following instructions

Criteria: Mark

Apparatus set up correctly, no assistance 2
        . . .with assistance 1
Adds copper oxide.. 1
Weighs test tube with copper oxide...................... 1
Re-assembles apparatus correctly 1
Lights gas and adjusts.. 1
                        Maximum mark 6

Mark Obtained	Skill Level	
	NEA/LEAG/NISEC	MEG/SEG/WJEC
5 and 6	5	3
4	4	2
3	3	
2	2	1
1	1	

Laboratory technician's notes

The apparatus and materials required for this assignment are:

  bunsen burner and mat
  safety glasses/goggles
  an accurate top-pan balance
  spatula
  bung with delivery tube (as shown)
  clamp stand with the clamp in the lowered position (as shown)
  pre-weighed amount of copper oxide on paper (about 4 or 5 grams)
  test tube with hole (0.25 cm diameter) 2 cm from end (as shown)

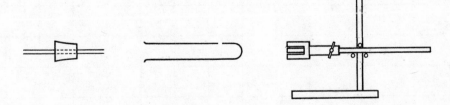

This apparatus has to be ready and in position before the assignment starts, giving teachers more time to make accurate assessments. You should try to stay in the laboratory during this assessment.

Skills check list – *Manipulation*

Pupil's Name	Raises clamp correctly by unscrewing	Refastens screw correctly	Clamp in safe position, teeth supporting	Places test tube in clamp correctly	Tightens up clamp correctly	Puts copper oxide into test tube without spillage	Weighs test tube without spillage	Correctly replaces tube with hole facing upwards	Sets gas correctly to small safe flame	Total Mark	Skill Level
	1		1	1	1	1	1	1	1		

Skills check list – *Following instructions*

Pupil's Name	Apparatus set up without assistance	Apparatus set up with assistance	Adds copper oxide	Weighs test tube with copper oxide	Re-assembles apparatus correctly	Lights gas and adjusts	Total Mark	Skill Level
	2	1	1	1	1	1		

Assignment 12
Investigating the rate of reaction between a metal and an acid

Teacher's notes

This is a teacher-demonstrated experiment because the whole group needs to have the same recordings, as required by some of the examination groups where collated results are needed for comparability of standards. Also, the concept of reaction rates is a difficult one anyway and is probably best explained by a teacher in the performance of a practical.

The reaction should be carried out using cleaned magnesium ribbon (about 0.1 g) and samples of dilute hydrochloric acid of different molarities (1.5M and 2.0M). About 15 to 20 ml of acid should be used. The yield of hydrogen should be no more than 60 ml.

Point out to pupils that you will read out results and they are expected to make their own recordings on the worksheets. However, it is useful if you also make notes for later use, to help less able pupils. Send the pupils to different areas of the room to make their own attempts at the graph work and its interpretation. Assistance should be given only when it is requested.

The skill *Recording observations* is more limited than usual as it does not involve the pupils *making* any observations. It is assessed by written performance, as are the skills *Presenting data* and *Drawing conclusions*. You will be able to assist pupils if necessary after the experiment is complete. Remember that lack of one skill should not prohibit a pupil from attempting another. For example, an inability to record observations should not stop the pupil drawing a graph, and an inability to draw a graph should not prevent pupils making a conclusion.

Marking schemes

Skill: Recording observations

Criteria:	Mark
Step 1(a), both acid strengths recorded correctly	1
Step 1(c), measurements of time and volume correct	1
Step 1(d), both temperature questions correct	1
Table is: correctly filled in	3
filled in with minor error	2
filled in with some errors	1
Maximum mark	6

Mark Obtained	Skill Level	
	NEA/LEAG/NISEC	MEG/SEG/WJEC
5 and 6	5	3
4	4	2
3	3	2
2	2	1
1	1	1

Skill: Presenting data

Criteria: Mark

Both axes correctly labelled and with units.................2
Most points plotted correctly....................................2
Both graphs reasonably drawn...............................1
Good use of scales...1
                                        Maximum mark...............6

Mark Obtained	Skill Level	
	NEA/LEAG/NISEC	MEG/SEG/WJEC
5 and 6	5	3
4	4	2
3	3	2
2	2	1
1	1	1

Skill: Drawing conclusions

Criteria: Mark

Step 7, reaction was faster as acid was stronger...............2
Step 8, correct sketch of graph 2
Step 10, temperature change could change the result........2
Step 11, increase in concentration increased the
        reaction rate (or similar)...2
        Points out error (e.g. syringe moves
        when bung is inserted)..2
                                        Maximum mark.................10

Mark Obtained	Skill Level	
	NEA/LEAG/NISEC	MEG/SEG/WJEC
9 and 10	5	3
7 and 8	4	2
5 and 6	3	2
3 and 4	2	1
1 and 2	1	1

Laboratory technician's notes

This is a demonstration experiment only. The apparatus should be ready and set up for the teacher to start the assignment. The apparatus and materials required are:

magnesium ribbon (clean)
1.5M hydrochloric acid (labelled)
2.0M hydrochloric acid (labelled)
Apparatus set up as shown in diagram on pupil's sheet:
   clamp containing boiling tube with delivery tube to syringe in another clamp
100 ml syringe
graph paper (A4 suggested)
stop clock

The apparatus should be set up so that all pupils can view the demonstration. If a syringe is unavailable then an inverted measuring cylinder can be used to collect the gas over water.

Assignment 13
Investigating the displacement of metals

Teacher's notes

It is possible for the whole group to carry out this assignment for the skill *Drawing conclusions*. This can be done by having a small group of pupils under examination conditions for the skill *Making and recording observations* with the rest of the group carrying out the practical as normal. At the end of the assignment, you could collate the findings in general and then the whole class could be placed under examination conditions to complete the exercise.

It is suggested that powdered forms of magnesium, zinc, iron and copper are used as the metals with any suitable salts.

Marking schemes

Skill: Making and recording observations

Criteria: *Mark*

There are 16 observations to be made and recorded in the table .... ½ each
Maximum mark 8

Mark Obtained	Skill Level	
	NEA/LEAG/NISEC	MEG/SEG/WJEC
8 and 7	5	3
6 and 5	4	2
4	3	2
3 and 2	2	1
1	1	1

Skill: Drawing conclusions

Criteria: *Mark*

Step 6 1
Step 7 1
Step 8 1
Correct conclusion, no assistance, suggests errors 3
Correct conclusion, but with minimum guidance,
    still suggests errors 2
Needs guidance to reach a conclusion 1
Maximum mark 6

Mark Obtained	Skill Level	
	NEA/LEAG/NISEC	MEG/SEG/WJEC
5 and 6	5	3
4	4	2
3	3	
2	2	1
1	1	

Laboratory technician's notes

Pupils will require the following apparatus and materials:

> metal salts in solution, e.g. magnesium chloride, zinc nitrate, iron(II) chloride, copper sulphate. These solutions should be about 1M in strength.
> metals in powder form: magnesium, zinc, iron and copper
> test tubes and rack
> spatula

The apparatus should be ready at the start of the assignment, with all the equipment and materials laid out ready for the pupils being assessed. The remainder of the group can collect their own equipment, but this should be placed as far away from the assessment area as possible, so as not to disturb the pupils being assessed.

Assignment 14
Designing an experiment to place a metal in the reactivity series

Teacher's notes

This is a design assignment. It is assessed by written performance therefore it can be used for the whole group. Some previous work on the reactivity of metals is essential for the success of this assignment, including consideration of displacement reactions, as in Assignment 13.

The assignment must be done under examination conditions, although there is no reason why pupils should not have access to notes or books.

Although the SEG does not require this skill to be assessed, you may find the assignment useful practice for pupils sitting the formal SEG examination at the end of the course.

The design of an experiment is not easy to assess because valid alternatives may arise. Therefore, in this assignment a criteria-related mark scheme is offered. Other design assignments which have specific marking schemes are offered elsewhere in this text.

Marking scheme

Skill: Designing an experiment

Mark Obtained	Skill Level	
	NEA/LEAG/NISEC	MEG/SEG/WJEC
Requires no guidance to produce any appropriate experiment	5	3
Requires minimum guidance to design an appropriate experiment	4	2
Requires guidance to design an appropriate experiment	3	2
Requires considerable guidance but is then able to proceed	2	1
Only with constant assistance can the design be carried out	1	1

Laboratory technician's notes

If the teacher decides to allow pupils to try out their designs, you should liaise with the teacher over the equipment and materials needed.

Assignment 15
Separating materials into conductors and non-conductors

Teacher's notes

This is a design assignment assessed by written performance, so it could be used for the whole group. Pupils will need some prior knowledge about conduction of electricity. This assignment could be done during a study of electrochemistry, although it is listed under metals and non-metals.

It would be helpful for pupils to carry out their experimental designs as soon after this assignment as possible, thus reinforcing what they have learned in the process.

Marking scheme

Skill: Designing an experiment

Criteria:	Mark
Draws an appropriate circuit	2
. . . after guidance	1
Indicates possible area of failure	1
Suggests using a known conductor to test circuit before use	1
Indicates correct/expected results	1
Suggests testing circuit after each test	1
Maximum mark	6

Mark Obtained	Skill Level	
	NEA/LEAG/NISEC	MEG/SEG/WJEC
5 and 6	5	3
4	4	2
3	3	2
2	2	1
1	1	1

*External assessment of this skill is not required by the SEG.

Laboratory technician's notes

If the teacher decides to follow up this assignment with a practical session, you should liaise over the equipment and materials needed.

Assignment 16
Investigating a mineral

Teacher's notes

This assignment can be carried out by the whole group with some pupils being observed more closely for assessing the skills *Manipulation* and *Following instructions*. Check lists are provided to assist with this part of the assignment (see pages 62 and 63). It should be possible to consider a full group assessment for *Making and recording observations* provided you can show it is the pupils' own work.

Marking schemes

Skill: Following instructions

Criteria:	Mark
Step 1 correct – limestone in tongs	1
Wears safety glasses	1
Step 3 correct – cools limestone	1
Step 4 correct – water/limestone	1
Step 5 correct – water/cool limestone	1
Step 6 correct – adds water	1
Maximum mark	6

Mark Obtained	Skill Level		
	NEA/LEAG/NISEC	MEG/SEG	WJEC
5 and 6	5	3	2
4	4	2	
3	3		1
2	2	1	
1	1		

Skill: Manipulation

Criteria:	Mark
Places limestone in tongs correctly, heats carefully	1
Places limestone into test tube, adds water, no spillage	1
Adds more water and pours, no spillage	1
Sets up filtration apparatus correctly	2
. . .but needs assistance	1
Carries out filtration with care	1
Maximum mark	6

Mark Obtained	Skill Level	
	NEA/LEAG/NISEC	MEG/SEG/WJEC
5 and 6	5	3
4	4	2
3	3	
2	2	1
1	1	

Skill: Making and recording observations

Criteria	Mark
Step 2 correct.............................. 1	
Step 3 correct.............................. 1	
Step 4 correct.............................. 1	
Step 5 correct.............................. 1	
Step 9 correct.............................. 1	
Step 10 correct 1	

Maximum mark ... 6

Mark Obtained	Skill Level	
	NEA/LEAG/NISEC	MEG/SEG/WJEC
5 and 6	5	3
4	4	
3	3	2
2	2	
1	1	1

Laboratory technician's notes

Each pupil carrying out this assignment will require the following:

some limestone
tongs
pH papers
filter funnel
filter papers
2 boiling tubes and rack
beaker
glass tube suitable for blowing through
stirring rod
safety glasses/goggles
ceramic mat
bunsen burner

Some pupils may carry out this assignment without being assessed. These pupils can collect their own apparatus and materials. However, any pupils involved in the assessment will need their apparatus and materials ready in the assessment area at the start of the assignment.

Skills check list – *Following instructions*

Pupil's Name	Step 1 correct – limestone in tongs	Wears safety glasses	Step 3 correct – cools limestone	Step 4 correct – water/limestone	Step 5 correct – water/ cool limestone	Step 6 correct – adds water	Total Mark	Skill Level
	1	1	1	1	1	1		

Skills check list – *Manipulation*

Pupil's Name	Places limestone in tongs correctly, heats safely	Places limestone in test tube, adds water, no spillage	Adds more water and pours, no spillage	Sets up filtration apparatus with assistance	Sets up filtration apparatus without assistance	Carries out filtration with care	Total Mark	Skill Level
	1	1	1	2	1	1		

Assignment 17
Making electricity from chemicals

Teacher's notes

This assignment can be carried out as a group activity with individual pupils carrying out different parts of the experiment. For example, it should be possible for one pupil to be making the recordings and another carrying out the manipulation. Alternatively, one section of the class could be carrying out this experiment as they would under normal laboratory conditions and a small number of pupils could be carrying out the whole assignment under your supervision.

As this assignment is to do with the skills *Manipulation* and *Making and recording observations* pupils should not be penalised if they need an explanation of how to fill in their tables before the assignment starts. A check list for *Manipulation* is available on page 66.

Marking schemes

Skill: Making and recording observations

Criteria: *Mark*

There are six recordings to be made ....1 mark each
Maximum mark.................. 6

Mark Obtained	Skill Level	
	NEA/LEAG/NISEC	MEG/SEG/WJEC
5 and 6	5	3
4	4	2
3	3	2
2	2	1
1	1	1

Skill: Manipulation

Criteria: *Mark*

Places plates in beaker, positions insulating strip 1
Connects plates to wires correctly 1
Connects wires to voltmeter correctly........................... 1
Pours in salt solution without spillage 1
Disconnects the apparatus correctly and replaces plate 1
Maximum mark.................. 5

Mark Obtained	Skill Level	
	NEA/LEAG/NISEC	MEG/SEG/WJEC
5	5	3
4	4	
3	3	2
2	2	
1	1	1

Laboratory technician's note

Each pupil being assessed will require the following equipment and materials:

> an insulation strip (a piece of thin plywood or similar, to fit across a beaker as a method of keeping the plates separate)
> various named metal strips of approximately the same size (e.g. magnesium, copper, zinc, lead and iron or steel plates. They should be clean and free from grease. Further supplies of magnesium ribbon should be available as this does tend to disappear rather quickly!
> a suitable voltmeter (0–5 V)
> strong salt solution (about 2M)
> a 250 ml beaker
> connecting leads with crocodile clips on one end, to fit voltmeter on the other end

Make sure that all the equipment is set up at the start of the session for those pupils being assessed. If the teacher wishes to extend the assignment to the whole group, extra equipment should be available for pupils to collect.

You may be needed in the laboratory to help the teacher during this assignment.

Skills check list – *Manipulation*

Pupil's Name	Places plates in beaker, positions insulating strip	Connects plates to wires correctly	Connects wires to voltmeter correctly	Pours in salt solution without spillage	Disconnects apparatus correctly and replaces plate		Total Mark	Skill Level
	1	1	1	1	1			

Assignment 18
Finding the freezing point of water

Teacher's notes

This assignment provides an opportunity for you to assess a number of skills at one time, and to use the results of the experiment to introduce a number of important concepts relevant to your pupils' studies. It should be possible for all pupils to produce a graph of collated results. Some pupils can have their skill at *Following instructions* assessed and a skills check list is provided on page 70 to facilitate this. Also, it is possible for a large group of pupils to have their ability at *Presenting data in tables and graphs* assessed by their written performance.

If a pupil cannot produce his own table for presenting data this should in no way affect any other skill area. A table is provided on page 71 for pupils who have problems in demonstrating this skill. This should be photocopied and distributed only when pupils show clearly that they cannot proceed beyond Step 1 in the assignment. You should provide all pupils with graph paper for Step 7.

This assignment can be used to introduce the idea that it is possible to measure the freezing point of water in the laboratory by a chemical method, but perhaps more importantly the idea that heat is used up when substances dissolve and this is the reason why freezing points can be lowered by adding ionic compounds. This work could be extended into the solvation of ions.

Marking schemes

Skill: Following instructions

Criteria:	Mark
Collects all apparatus, without assistance	2
. . . with assistance	1
Sets up apparatus as shown	1
Step 4, places thermometer correctly	1
Step 5, adds ammonium chloride correctly	1
Step 6, takes readings and records them correctly	1
Maximum mark	6

Mark Obtained	Skill Level		
	NEA/LEAG/NISEC	MEG/SEG	WJEC
5 and 6	5	3	2
4	4	2	1
3	3		
2	2	1	
1	1		

Skill: Presenting data (table)

Criteria:	Mark
Single table	1
Table is well-presented and tidy	1
Table uses columns	1
Columns are correctly titled	1
Table contains reasonable set of entries	1
Table is logical	1

Maximum mark 6

Mark Obtained	Skill Level	
	NEA/LEAG/NISEC	MEG/SEG/WJEC
5 and 6	5	3
4	4	2
3	3	
2	2	1
1	1	

Skill: Presenting data (graph)

Criteria:	Mark
Axes are correctly labelled, with units	2
Most points are in correct place	2
Graph is of a reasonable shape	2

Maximum mark 6

Mark Obtained	Skill Level	
	NEA/LEAG/NISEC	MEG/SEG/WJEC
5 and 6	5	3
4	4	2
3	3	
2	2	1
1	1	

Laboratory technician's notes

Pupils being assessed in this assignment will require the following:

distilled or de-ionised water
ice
boiling tube
ammonium chloride (about 5 g)
thermometer (−10°C to 110°C)
stop clock
250 ml beaker
spatula
graph paper

All pupils should collect their apparatus for this assignment, including those being assessed.

Skills check list – *Following instructions*

Pupil's Name	Collects all apparatus without assistance	Collects all apparatus with assistance	Sets up apparatus as shown	Step 4, places thermometer correctly	Step 5, adds ammonium chloride correctly	Step 6, takes readings, records them correctly	Total Mark	Skill Level
	2	1	1	1	1	1		

Name..Group...............................Date.............

Use the following table to fill in your results for steps 2–7 and to complete this assignment:

Time (½ min)	Temperature (°C)	Description of distilled water

If you need to continue your table, use the other side of this sheet.

Assignment 19
Determining the boiling point of a liquid

Teacher's notes

This assignment requires a fairly high level of manipulative skill, e.g. fastening, setting levels, pouring. As an assessment it may provide discrimination for the more able pupils, although it is expected that below-average pupils will attempt the assignment and need assistance.

The two skills *Manipulation* and *Following instructions* should not be confused. It is possible to follow a set of instructions correctly with poor manipulative skills. Check lists for these skills are on pages 74 and 75.

This assignment can be carried out during the study of solids, liquids and gases.

The liquid you choose must have a reasonably low boiling point. However, as the choice is likely to be a hydrocarbon, and therefore highly inflammable, it is essential that Step 7 (turning off bunsen burners) is observed by all pupils.

At Step 10 you need to provide data pages and fill in the table on the worksheet with suitable data before photocopying. This would include names of suitable liquids and their boiling points, relevant to what you have available in the laboratory.

Marking schemes

Skill: Manipulation

Criteria: *Mark*

Step 1, fastens thermometer correctly with no assistance 1
Step 2, fastens tube correctly with no assistance 1
Step 3, carefully clamps test tube without trapping the thermometer 1
Step 4, manipulates clamp to right level 1
Step 4, manipulates clamp to facilitate placing beaker 1
Step 6, lights bunsen burner safely, adjusts flame 1
Maximum mark 6

Mark Obtained	Skill Level	
	NEA/LEAG/NISEC	MEG/SEG/WJEC
5 and 6	5	3
4	4	2
3	3	2
2	2	1
1	1	1

Skill: Following instructions

       Criteria: *Mark*

       Steps 1 to 6 are all practical instructions
          and for each step correctly completed ....1 mark each
                       Maximum mark 6

Mark Obtained	Skill Level		
	NEA/LEAG/NISEC	MEG/SEG	WJEC
5 and 6	5	3	2
4	4	2	
3	3		
2	2	1	1
1	1		

Laboratory technician's notes

All pupils being assessed will require the following apparatus:

    suitable liquid – such as ethanol
    test tube
    thermometer (−10 to 110°C)
    small piece of capillary tube, sealed at one end, about half the length of the test tube
    elastic bands
    bunsen burner
    tripod
    gauze
    mat

This apparatus should be ready at a particular point in the laboratory for each pupil. These points should be close enough to assist the teacher observing the pupils, but far enough apart to ensure pupils produce their own work.

You should be present in the laboratory for the assessment if possible.

Skills check list – *Manipulation*

Pupil's Name	Step 1, fastens thermo-meter with no assistance	Step 2, fastens tube, no assistance	Step 3, clamps test tube without trapping thermometer	Step 4, manipulates clamp to right level	Step 4, manipulates clamp to facilitate placing beaker	Step 6, lights bunsen safely, adjusts flame	Total Mark	Skill Level
	1	1	1	1	1	1		

Skills check list – *Following instructions*

Pupil's Name	Step 1 correct 1	Step 2 correct 1	Step 3 correct 1	Step 4 correct 1	Step 5 correct 1	Step 6 correct 1	Total Mark	Skill Level

Assignment 20
Using melting points to find impurities

Teacher's notes

This assignment is to do with the pupil's ability to use a given experiment to produce a method of presenting information in a table and a graph. These skills require a high level of ability. You are entitled to help your pupils reach their level of ability, but to avoid possible problems you should have photocopies of the ready-made table on page 78 available so that pupils who cannot present their own table can attempt the graph.

It is suggested that the apparatus shown in the laboratory technician's notes is used and that the experiments (i.e. for A and B) are performed at the same time. It would be useful to have both test tubes in the same beaker to avoid confusion over different starting temperatures for the cooling curves.

Pupils should be familiar with the idea that pure substances have a fixed melting point, but when the same material is impure it may melt over a range of temperatures. Also, this assignment is a useful way of introducing the idea that a fixed melting point is a criterion for purity.

You should perform this experiment in full view of all the pupils, but inform them that they must work on their own. After taking readings, pupils would be dispersed around the room to produce their own graphs. Check that pupils have the correct readings; their ability to draw a graph should not be penalised by an inability to record information in a table.

Marking schemes

Skill: Presenting data (table)

Criteria:	Mark
Draws a logical table without any assistance	2
...with assistance	1
Table is neat and tidy	1
Recordings are correctly titled	1
Information is easy to obtain	1
Uses a single table	1
Maximum mark	6

Mark Obtained	Skill Level	
	NEA/LEAG/NISEC	MEG/SEG/WJEC
5 and 6	5	3
4	4	2
3	3	
2	2	1
1	1	

Skill: Presenting data (graph)

Criteria:	Mark
Both axes correctly labelled	2 (1 each)
Good use of axes by scale	2 (1 each)
Both sets of points correctly plotted	2 (1 each)
Graphs are reasonably constructed	2 (1 each)
Melting points are labelled	2 (1 each)
Maximum mark	10

Mark Obtained	Skill Level	
	NEA/LEAG/NISEC	MEG/SEG/WJEC
9 and 10	5	3
7 and 8	4	2
5 and 6	3	2
3 and 4	2	1
1 and 2	1	1

Laboratory technician's notes

The teacher will require the following apparatus set up in a place visible to all pupils:

2 thermometers (0 to 110°C)	2 tripods
*test tube with solid	gauze
*test tube with impure solid	2 clamp stands
beaker	bunsen burner and mat

*The solid to be used should be the same in both test tubes but with one made impure by adding a small amount of, for example, wax. Liaise with the teacher over a suitable solid to use for this experiment. Naphthalene should *not* be need.

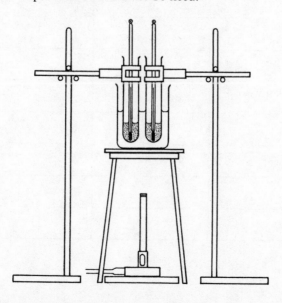

Assignment 20
Supplement to worksheet

Name.. Group...............................Date............

Fill in the following table as your teacher carries out the experiment:

Time (½ min)	Temp°C (A)	Temp°C (B)

Assignment 21
Finding the melting point of a solid

Teacher's notes

This assignment can involve the whole class, even though only some pupils are assessed for the skill *Following instructions* and *Manipulation* (skills check lists on pages 82 and 83). This will include selecting apparatus and setting it up. Other pupils could observe this and a discussion could follow the assessment. Involving pupils in such friendly discussions is more effective than just telling a pupil where they went wrong.

When the experiment has eventually reached step 8 the class can revert to normal laboratory conditions until the skill *Presenting data in graphs* is to be assessed. At this point pupils should be dispersed around the room to complete their own graphs. Check that all pupils have more or less the same results. If not, then a set of collated results could be provided for all pupils to make a second attempt at this skill.

Marking schemes

Skill: Following instructions

Criteria:	Mark
Selects tripod and gauze	1
bunsen burner	1
beaker and test tube with solid	1
clamp stand	1
thermometer	1
timer	1
Step 2, clamp in correct position	1
Step 3, full blue flame	1
Step 4, solid melts	1
Step 6, apparatus changed correctly	1
Step 7, starts stop clock	1
Step 8, safe position	1
Maximum mark	12

Mark Obtained	Skill Level		
	NEA/LEAG/NISEC	SEG/MEG	WJEC
10, 11 and 12	5	3	2
8 and 9	4	2	
6 and 7	3		1
4 and 5	2	1	
1, 2 and 3	1		

Skill: Manipulation

Criteria:	Mark
Step 2, sets up apparatus correctly, needs	
no assistance 3	
.....needs minimal assistance 2	
.....needs assistance 1	
Step 3, carefully lights bunsen to correct flame 1	
Step 6, carefully removes apparatus 1	
Step 8, manipulates clamp to safe position 1	
Maximum mark................ 6	

Mark Obtained	Skill Level	
	NEA/LEAG/NISEC	SEG/MEG/WJEC
5 and 6	5	3
4	4	2
3	3	
2	2	1
1	1	

Skill: Presenting data

Criteria:	Mark
Correctly labelled axes .... 2 (1 each)	
Plotted results reflect pupils' results 2	
Attempted to draw reasonable curve 2	
...dot-to-dot rather than curve 1	
Maximum mark............... 6	

Mark Obtained	Skill Level	
	NEA/LEAG/NISEC	SEG/MEG/WJEC
5 and 6	5	3
4	4	2
3	3	
2	2	1
1	1	

Laboratory technician's notes

This assignment requires pupils to collect their own apparatus, which should be in the laboratory ready for the start of the assessment. Do not set up the apparatus in any way, as you would be for other assignments.

Each pupil will be required to collect the following:

> thermometer (0 to 110°C)
> test tube containing a solid (melting point about 80°C)
> beaker
> tripod
> gauze
> clamp stand and clamp
> bunsen burner and mat
> stop clock

Skills check list – *Following instructions*

Pupil's Name	Selects apparatus (1 mark per item)	Step 2, clamp in correct position	Step 3, full blue flame	Step 4, solid melts	Step 6, apparatus changed correctly	Step 7, starts stop clock	Step 8, safe position	Total Mark	Skill Level
	6	1	1	1	1	1	1		

Skills check list – *Manipulation*

Pupil's Name	Sets up apparatus without assistance	… with minimal assistance	… with assistance	Step 3, carefully lights bunsen to correct flame	Step 6, carefully removes apparatus	Step 8, manipulates clamp to safe position	Total Mark	Skill Level
	3	2	1	1	1	1		

Assignment 22
Investigating fermentation

Teacher's notes

This is a written assignment, aimed at a large group of pupils. It should be done only after a study of fermentation and when pupils are familiar with the idea of distillation.

It is always difficult to suggest a marking scheme for experimental design; the one here is given only as a guide. It may be that pupils will design a totally different experiment which may still work. For instance, it is possible to use a hydrometer to test the results of fermentation, but this does not require much ingenuity on the part of the pupil. To overcome this, it may be better to suggest to pupils that they use the apparatus listed on the sheet before trying alternatives.

Reference should be made to the general criteria listed on page 17.

The Southern Examining Group assesses the skill *Designing an experiment* by examination. However, if you are teaching the SEG syllabus, you can still use these assignments to prepare pupils for this part of their formal assessment. You may wish to set this exercise for homework.

Marking scheme

Skill: Designing an experiment

Criteria:	Mark
Weighs out sugar	1
Same amount of water used in each sample	1
Adds yeast, about same amount, uses cotton wool plug	1
Several test tubes kept in places with different controlled or known temperatures	1
Suggests a time limit	1
Suggests a sensible method for measuring the amount of alcohol produced, e.g. distillation/hydrometer	1
Maximum mark	6

Mark Obtained	Skill Level	
	NEA/LEAG/NISEC	MEG/SEG/WJEC
5 and 6	5	3
4	4	2
3	3	
2	2	1
1	1	

*External assessment of this skill is not required by the SEG.

Laboratory technician's notes

If the teacher decides to follow up this written assignment with a practical session, to try out the pupils' designs, then you should liaise with the teacher over equipment and materials required.

Assignment 23
Using chromatography to identify an impurity

Teacher's notes

As this is a written experiment it is possible to do this assessment with the whole group. Pupils will need some prior knowledge and practical awareness of chromatography. When the assignment is completed it should be followed up with a real laboratory situation if possible, so that pupils are able to try out their own ideas.

The marking scheme given below is for guidance only. It is possible that pupils will design experiments which are different from the one 'expected', so you should keep an open mind when assessing design experiments such as this.

Although the Southern Examining Group does not require external assessment of this skill, you may find the assignment useful practice for pupils sitting the formal SEG examinations at the end of the course.

Marking scheme

Skill: Designing an experiment

Criteria:	Mark
Suggests using a standard sample for purposes of comparison	2
Indicates testing samples over the ten days	1
Indicates any correct procedure for testing the pens for the dye...without assistance	3
...with minimal assistance	2
...with much assistance	1
Maximum mark	6

Mark Obtained	Skill Level	
	NEA/LEAG/NISEC	MEG/WJEC (SEG)*
5 and 6	5	3
4	4	2
3	3	2
2	2	1
1	1	1

*External assessment of this skill is not required by the SEG.

Laboratory technician's notes

You should liaise with the teacher over what equipment and materials are necessary for the follow-up session to this assignment.

Assignment 24
Investigating river pollution

Teacher's notes

This written assignment can be used for a whole group. Some previous work on acids, alkalis and indicators is required. The task would benefit from being carried out during any study on pollution, and backed-up with practical investigation afterwards. For example "polluted" water samples could be provided so that pupils can carry out their own designs after the assignment. This may mean that some designs will have to be amended, as part of the teaching process involved in designing experiments.

Although the Southern Examining Group does not require you to assess this skill, you may find the assignment useful preparation for the SEG examination.

Marking scheme

Skill: Designing an experiment

Criteria:	Mark
Suggests sampling upstream	2
Samples are measured by volume	1
Records time of day and date	1
Outlines any sensible method of testing	1
Outlines any sensible method of keeping results	1
Maximum mark	6

Mark Obtained	Skill Level	
	NEA/LEAG/NISEC	MEG/WJEC (SEG)*
5 and 6	5	3
4	4	2
3	3	2
2	2	1
1	1	1

*External assessment of this skill is not required by the SEG.

Assignment 25
Investigating the pollution near a sulphuric acid plant

Teacher's notes

This is a design assignment. Since it is assessed by written performance it can be used for the whole group, but you should follow up the pupils' efforts with a reliable experiment if possible. The assignment should form part of a study on pollution or titrations, or both.

As in any design assignment, pupils may produce many correct possibilities, so you may have to refer to the general marking schemes in Chapter 3. You should direct pupils to notice the items listed, and encourage them to use these items in their design. The mark scheme below offers guidelines to this effect.

Designing experiments is not part of the external assessment for the SEG, but you will hopefully find this assignment useful as practice for the internal assessment.

Marking scheme

Skill: Designing an experiment

Criteria:	*Mark*
Indicates titration method	1
Suggests that samples taken must be of same volume	1
Uses an indicator	1
Mentions neutral point	1
Suggests sampling over a range of distances	1
Suggests sampling from lake as a reference	1
Maximum mark	6

Mark Obtained	Skill Level	
	NEA/LEAG/NISEC	MEG/WJEC (SEG)*
5 and 6	5	3
4	4	2
3	3	2
2	2	1
1	1	1

*External assessment of this skill is not required by the SEG.

Laboratory technician's notes

If the teacher decides to follow up this assignment with an experiment, you should liaise with the teacher over what equipment and materials are required.

Assignment 26
Investigating the rate of a reaction

Teacher's notes

This is a design assignment. Since it is assessed by written performance it can be used for the whole group. However, it should be followed up at some stage with an experiment, to show pupils how it can be done.

The SEG does not require this skill to be assessed as part of the external assessment procedure. However, pupils will require practice before having the assessment in the formal examination at the end of the course.

Marking scheme

Skill: Designing an experiment

Criteria:	Mark
Indicates weighing out exact and identical amounts of calcium carbonate	1
Uses measuring cylinder to measure out identical amounts of acid	1
Suggests wearing goggles/safety glasses	1
Puts acid in funnel	1
Puts carbonate in flask	1
Adds all acid at one go	1
...and starts clock	1
Indicates time to be recorded	1
Indicates volume to be measured	1
Draws a reasonable table for recordings	1
Logical sequence to the experiment	1
Suggests possible sources of errors	1
Maximum mark	12

Mark Obtained	Skill Level	
	NEA/LEAG/NISEC	MEG/WJEC (SEG)*
10, 11 and 12	5	3
8 and 9	4	2
6 and 7	3	2
4 and 5	2	1
1, 2 and 3	1	1

*External assessment of this skill is not required by the SEG.

Laboratory technician's notes

If the teacher decides to follow up this assignment with an experiment, you should liaise with the teacher over the equipment and materials required.

Assignment 27
Investigating the rate of a reaction by loss of mass

Teacher's notes

This is a teacher-demonstrated experiment assessed by written performance and can therefore be used to assess the whole group. It may well be useful for pupils who have previously missed this type of assessment.

The amounts of calcium carbonate used and the strengths of acid are very important. The acids should vary from 1M to 3M. Pupils should know exactly what these strengths are. The amount of calcium carbonate should be determined by experimental procedure rather than calculation. In practice, the amount of acid should be about 50 ml and the amount of calcium carbonate should be 10 g. These amounts should be fixed for all three experiments so that results are comparable.

It is possible that pupils will miss the first few readings during each experiment because the initial reaction is very fast. You should explain this to pupils and try to give them accurate readings yourself.

Marking scheme

Skill: Presenting data in graphs

Criteria:	Mark
Both axes correctly labelled	2
Good use of scales on each axis	1
Most points correct	2
...reasonable number correct	1
Good shape to graph	1
Maximum mark	6

Mark Obtained	Skill Level	
	NEA/LEAG/NISEC	MEG/SEG/WJEC
5 and 6	5	3
4	4	2
3	3	2
2	2	1
1	1	1

Laboratory technician's notes

The teacher will require the following apparatus and materials set up in a position visible to a large group of pupils:

1, 2 and 3 M hydrochloric acid
calcium carbonate powder
accurate balance (preferably electronic)

3 conical flasks
stop clock

Assignment 28
Preparing oxygen to investigate a catalyst

Teacher's notes

As this is a design assignment pupils are not actually required to carry out their experiments. However, a demonstration of the corrected experiment should be performed, as part of the learning process. This assignment would be usefully set during work on oxygen and/or catalysts.

The marking scheme given below is for *guidelines* only, as pupils will possibly produce unusual experiments. Reference should be made to the general criteria in Chapter 3.

The SEG does not require this skill to be assessed as part of its external assessment procedure. However, you should hopefully find this assignment useful practice for your pupils.

Marking scheme

Skill: Designing an experiment

Criteria:	Mark
Correct use of apparatus	1
Uses hydrogen peroxide on its own (no catalyst)	1
Uses same amount of hydrogen peroxide each time	1
Suggests timing the amount of gas liberated	1
Suggests comparing time to find best catalyst	1
Suggests catalyst should be unchanged	1
Maximum mark	6

Mark Obtained	Skill Level	
	NEA/LEAG/NISEC	MEG/WJEC (SEG)*
5 and 6	5	3
4	4	2
3	3	2
2	2	1
1	1	1

*External assessment of this skill is not required by the SEG.

Laboratory technician's notes

The teacher will let you know what equipment and materials are required for the demonstration.

Assignment 29
Investigating gases

Teacher's notes

This assignment is meant to follow work already carried out in the laboratory. It would most usefully be carried out during a study of the air and should involve some knowledge of liquifaction. Pupils should be allowed to use notes and textbooks during this assignment.

An assignment such as this could be of use when assessing pupils with genuine absence problems or those who, for whatever reason, have not been able to keep up with their assessments.

Marking scheme

Skill: Drawing conclusions

Criteria:	Mark
Question 1	3
Question 2	1
Question 3	1
Question 4	1
Question 5	1
Question 6	1
Question 7...improvement	1
...error	1
Maximum mark	10

Mark Obtained	Skill Level	
	NEA/LEAG/NISEC	MEG/SEG/WJEC
9 and 10	5	3
7 and 8	4	2
5 and 6	3	2
3 and 4	2	1
1 and 2	1	1

Laboratory technician's notes

You are not needed for this assignment.

Assignment 30
Preparing carbon dioxide gas

Teacher's notes

This assignment is intended for a small number of pupils to assess the skills *Following instructions* and *Manipulation*. This will probably mean that the remainder of the group will need an alternative assignment, such as a design skill's test. The whole group could then be placed under examination conditions. The whole class could eventually be brought together to perform the investigation part of this assignment.

You are expected to carry out part 7(f) yourself. This part of the assignment should demonstrate that carbon dioxide is heavier/more dense than air. If you can collect a gas jar of carbon dioxide, this can then be 'poured' onto a burning candle at the bottom of another gas jar. This is shown in the diagram below.

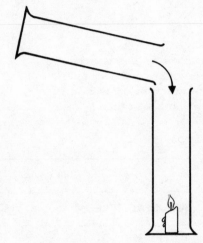

The skill *Drawing conclusions* can be assessed by written performance after the other skills have been assessed. All pupils should be given a copy of the second part of the worksheet, so that the conclusions they make are based on comparable information. Pupils should still make their own observations, but use the worksheet to overcome mistakes they might have made earlier.

Skills check lists are provided for *Following instructions* and *Manipulation* on pages 96 and 97.

Marking schemes

Skill: Following instructions

Criteria:	Mark
Selects apparatus required	2
Sets up apparatus as shown	2
Practises inverting test tube	2
Maximum mark	6

Mark Obtained	Skill Level		
	NEA/LEAG/NISEC	SEG/MEG	WJEC
5 and 6	5	3	2
4	4	2	
3	3		1
2	2	1	
1	1		

Skill: Manipulation

Criteria: Mark

Clamps test tube at mouth correctly............... 1
Inserts delivery tube correctly 1
Fills tub with water, no spillage 1
Correct inversion of test tube....................... 2
                                        Maximum mark 5

Mark Obtained	Skill Level	
	NEA/LEAG/NISEC	MEG/SEG/WJEC
5	5	3
4	4	2
3	3	
2	2	
1	1	1

Skill: Drawing conclusions

Criteria: Mark

Colourless 1
Soluble 1
Does not support burning.................... 1
Acidic.. 1
Heavier than air, or more dense 1
                                        Maximum mark 5

Mark Obtained	Skill Level	
	NEA/LEAG/NISEC	SEG/MEG/WJEC
5	5	3
4	4	
3	3	2
2	2	
1	1	1

Laboratory technician's notes

Each pupil being assessed will require the following items:

> hydrochloric acid (2M)
> 5 test tubes and bungs
> test tube rack
> delivery tube to fit
> boiling tube
> safety glasses/goggles
> wooden splint
> limewater
> beaker (or margarine tub) to act as a trough
> indicator
> calcium carbonate
> spatula
> clamp stand
> a small test tube to fit inside a boiling tube as shown

You should remain in the laboratory for this assignment if possible, to assist the teacher.

Skills check list – *Following instructions*

Pupil's Name	Selects apparatus as required	Sets up apparatus as shown	Practises inverting test tube	Total Mark	Skill Level
	2	2	2		

Skills check list – *Manipulation*

Pupil's Name	Clamps test tube at mouth correctly	Inserts delivery tube correctly	Fills tube with water, no spillage	Correct inversion of test tube	Total Mark	Skill Level
	1	1	1	2		

Assignment 31
Investigating burning

Teacher's notes

This assignment is intended for the whole group to perform, although only a small group of pupils will be able to have their individual skills assessed. *Following instructions* and *Manipulation* can be assessed only by direct observation by the teacher (skills check lists on pages 100 and 101). The skill *Making and recording observations* can be assessed by written performance.

The materials used for burning should be organic, e.g. wood, paper, plastic, oil, candle and wool.

Marking schemes

Skill: Following instructions

Criteria: *Mark*

Step 2, arranges spoon as instructed............ 1
Step 3 to Step 7 as instructed ......1 mark each
               Maximum mark......... 6

Mark Obtained	Skill Level		
	NEA/LEAG/NISEC	MEG/SEG	WJEC
5 and 6	5	3	2
4	4	2	2
3	3	2	1
2	2	1	1
1	1	1	1

Skill: Manipulation

Criteria: *Mark*

Arranges spoon correctly... 1
Adds material and ignites it
  correctly and safely, wearing goggles......................... 1
Lowers burning material into gas jar safely 1
Removes spoon and places cover slip on jar correctly....... 1
Pours in some limewater without spilling and with
  correct procedure with cover slip 1
Shakes contents together, holds top on........................ 1
               Maximum mark.................. 6

Mark Obtained	Skill Level	
	NEA/LEAG/NISEC	MEG/SEG/WJEC
5 and 6	5	3
4	4	2
3	3	
2	2	1
1	1	

Skill: Making and recording observations

Criteria: *Mark*

There are six observations to be made
under the title 'What happened with the
limewater?'..1 mark each
Maximum mark...................... 6

Mark Obtained	Skill Level	
	NEA/LEAG/NISEC	SEG/MEG/WJEC
5 and 6	5	3
4	4	2
3	3	
2	2	1
1	1	

Laboratory technician's notes

The apparatus and materials listed below will be required by all pupils being assessed, and
possibly by the remainder of the group for a normal practical. *All* pupils are asked to collect
the following items:

limewater
various materials for burning, e.g. wood, wool, plastic (care),
    oil, candle or animal fat, leaves, paper
bunsen burner and mat
deflagrating spoon
gas jar and cover slip
safety glasses/goggles

Skills check list – *Following instructions*

Pupil's Name	Step 2, arranges spoon correctly 1	Step 3 correct 1	Step 4 correct 1	Step 5 correct 1	Step 6 correct 1	Step 7 correct 1	Total Mark	Skill Level

Skills check list – *Manipulation*

Pupil's Name	Arranges spoon correctly	Adds material and ignites it safely, wearing goggles	Lowers burning material into gas jar safely	Removes spoon and places cover slip on jar correctly	Pours limewater without spillage, correct use of cover slip	Shakes contents, holding top on, safely	Total Mark	Skill Level
	1	1	1	1	1	1		

Assignment 32
Investigating the hardness of water

Teacher's notes

This is a design assignment. It is assessed by written performance and can therefore be used for the whole group. The results can also be used for moderation, particularly within a school across a number of groups.

As with any design assignment the marking scheme remains vague, and can be used only for the suggested experiment. Pupils may come up with valid alternatives, in which case you should adhere to the general guidelines in Chapter 3.

Also, it is recommended that Assignment 33 is used to follow this one. This will assist in the learning process.

Although the SEG does not require the skill to be assessed externally, you should still find the exercise useful practice.

Marking scheme

Skill: Designing an experiment

Criteria:	Mark
Uses soap as a test for hardness	2
Uses measured amount of soap for comparison	1
Uses same amount of water each time	1
Uses distilled water as the standard	2
Maximum mark	6

Mark Obtained	Skill Level	
	NEA/LEAG/NISEC	MEG/WJEC (SEG)*
5 and 6	5	3
4	4	2
3	3	2
2	2	1
1	1	1

*External assessment of this skill is not required by the SEG.

Laboratory technician's notes

If the teacher decides to follow up this assignment with a practical session, you should liaise with the teacher over the items needed.

Assignment 33
Testing the hardness of various waters

Teacher's notes

It should be possible to carry out this assignment with a large group of pupils provided the equipment is available. Pupils must work on their own for the skill *Making and recording observations*. You should check that the pupils' observations are valid before going on to the *Drawing conclusions* work. Both these skills can be assessed by written performance.

The water samples provided can be: calcium sulphate, magnesium sulphate and calcium hydrogen carbonate.

Marking schemes

Skill: Making and recording observations

Criteria:	Mark
Step 2...correct level	1
Step 3...correct level	1
Check any four results in the table	1 mark each
Maximum mark	6

Mark Obtained	Skill Level	
	NEA/LEAG/NISEC	MEG/SEG/WJEC
5 and 6	5	3
4	4	2
3	3	2
2	2	1
1	1	1

Skill: Drawing conclusions

Criteria:	Mark
Step 1 – correct inference	1
In **Conclusions** section:	
1. Correct answer	1
3. Correct response to 'a lot less soap'	1
4(a) Correct answer	1
4(b) Correct answer	1
4(c) Correct answer	1
Maximum mark	6

Mark Obtained	Skill Level	
	NEA/LEAG/NISEC	MEG/SEG/WJEC
5 and 6	5	3
4	4	2
3	3	
2	2	1
1	1	

Laboratory technician's notes

The following equipment and materials should be available for each pupil taking part in the assignment:

    small measuring cylinder
    large measuring cylinder
    boiling tubes and bungs (preferably 7 each)
    boiling tube rack
    soap solution
    distilled water in a labelled beaker
    three labelled water samples A, B and C (for example, distilled water containing trace
      amounts of calcium sulphate, magnesium sulphate and calcium hydrogen carbonate).
    the same three water samples A, B and C *boiled*.

Assignment 34
Testing water for the presence of sulphates

Teacher's notes

This assignment could be carried out by the whole group but only a few pupils could be observed for the skill *Following instructions*. As the apparatus is so simple all pupils should be able to make their own individual set of recordings. The skill *Making and recording observations* can be assessed by written performance.

Pupils may find it worthwhile providing their own water from various sources for the whole group to analyse (e.g. rain water, pond water, river water, sea water, tap water, etc.) A total of six water samples should be presented for analysis. These should be tested before the assignment is started. It may be necessary to 'doctor' the samples to give a wide variety of results, but you should avoid this if possible.

Pupils should be made aware of what they are looking for, e.g. the precipitate which indicates a positive test.

A skills check list is provided on page 107 for assessing the skill *Following instructions*.

Marking schemes

Skill: Following instructions

Criteria:	Mark
Carries out sulphate test as instructed	
...without any assistance	3
...with small amount of assistance	2
...with much assistance	1
Carries out chloride test as instructed	
...without assistance	3
...with a small amount of assistance	2
...with much assistance	1
Maximum mark	6

Mark Obtained	Skill Level		
	NEA/LEAG/NISEC	MEG/SEG	WJEC
5 and 6	5	3	2*
4	4	2	
3	3		1
2	2	1	
1	1		

*The pupil carries out either set of instructions without assistance.

Skill: Making and recording observations

Criteria:	*Mark*
There are 12 observations to be made...............	1 mark each
Maximum mark	12

Mark Obtained	Skill Level	
	NEA/LEAG/NISEC	MEG/SEG/WJEC
10, 11 and 12	5	3
8 and 9	4	2
6 and 7	3	
4 and 5	2	1
1, 2 and 3	1	

Laboratory technician's notes

The following items should be available for each pupil doing the assignment:

> barium chloride solution
> silver nitrate solution
> hydrochloric acid
> nitric acid
> test tubes in rack (as many as can be spared)
> goggles/safety glasses

Pupils will be encouraged by the teacher to bring in their own water samples. If this is not possible, you should liaise with the teacher over the six water samples required.

Skills check list – *Following instructions*

Pupil's Name	Performs sulphate test ... without assistance	... with minimal assistance	... with much assistance	Performs chloride test ... without assistance	... with minimal assistance	... with much assistance	Total Mark	Skill Level
	3	2	1	3	2	1		

Assignment 35
Investigating rusting

Teacher's notes

This is a design assignment. It is assessed by written performance and can therefore be used for the full group. All pupils would benefit from their ideas being put to the test, as part of the learning process. The assignment should be set during a study of either the air or rusting, or perhaps during a study of iron.

It would be very useful if car bodywork could be obtained from a scrap yard. It would have to have any rust protection removed from one side.

The SEG makes no requirement for the design of experiments as part of their external assessment, but you should still find this assignment useful as a preparation for formal examinations.

The following marking scheme is for guidance only. Pupils may come up with viable alternatives, in which case you should refer to the general criteria presented in Chapter 3.

Marking scheme

Skill: Designing an experiment

Criteria:	Mark
Suggests leaving plate in damp place	1
Suggests leaving plate in damp place with salt	1
Suggests leaving plate in very dry place	1
Suggests leaving plate in dry place after heating and covering with dry sand	1
Compares plates	1
Suggests standard, e.g. eliminates air/water/salt	1
Maximum mark	6

| Mark Obtained | Skill Level | |
	NEA/LEAG/NISEC	MEG/WJEC (SEG)*
5 and 6	5	3
4	4	2
3	3	2
2	2	1
1	1	1

*External assessment of this skill is not required by the SEG.

Laboratory technician's notes

If pupils are going to carry out their designs then the following items will be required:

    car bodywork – cut into small pieces about 1 cm square
    test tubes and rubber bungs
    salt
    water
    a dry store room

Real car bodywork should be used if possible. A trip to the scrap yard may be required. Pupils may also ask for other equipment to carry out their own particular designs, so you should liaise with the teacher before the practical session.

Assignment 36
Investigating the causes and prevention of rusting

Teacher's notes

This is a traditional experiment which can serve a number of purposes. It could probably be used to test almost any skill involved in this subject, and could be used to extend the scope of Assignment 35.

The skills *Following instructions* and *Manipulation* can be assessed only in a limited way; it could be argued that if the test tubes are set up correctly then each pupil must have carried out the instructions correctly. That is for you to judge. If your department has some hundred test tubes to spare then you could attempt this assessment for the whole group.

The skill *Making and recording observations* could also be a whole group assessment. The pupils could take turns looking at collections of apparatus which contain the correct results. This would mean the whole group would also be able to be assessed for their ability at *Drawing conclusions*. This assignment may prove useful for pupils with genuine absence problems, because it covers so many skills.

The assignment would best be carried out during a study of rusting. It could be extended to include other methods of preventing rusting. For example, the same experiment could be set up with iron coated in various protective materials, such as paint, grease, wax, oil or plastic.

Skills check lists are provided on pages 113 and 114 for the assessment of *Following instructions* and *Manipulation*. The skills *Making and recording observations* and *Drawing conclusions* can be made by the pupil's written performance.

Marking schemes

Skill: Following instructions

Criteria: *Mark*

For each of the instructions 2(a) to (f).....
...no assistance............................. 2 marks each
...with assistance ....1 mark each
Maximum mark 12

Mark Obtained	Skill Level		
	NEA/LEAG/NISEC	SEG/MEG	WJEC
11 and 12	5	3	2
9 and 10	4	2	
6, 7 and 8	3		1
4 and 5	2	1	
1, 2 and 3	1		

Skill: Manipulation

        Criteria: *Mark*

Step 2(a)   Heats correctly and safely 2
Step 2(b)   Places iron, adds water, no spills............................2
Step 2(c)   Sets up correctly...2
Step 2(d)   Sets up correctly...2
Step 2(e)   Sets up correctly, particularly water level..................2
Step 2(f)   Sets up correctly, salt water level 2
                                               Maximum mark 12

Any assistance given in each step, award 1 mark only

Mark Obtained	Skill Level	
	NEA/LEAG/NISEC	MEG/SEG/WJEC
11 and 12	5	3
9 and 10	4	2
6, 7 and 8	3	
4 and 5	2	1
1, 2 and 3	1	

Skill: Making and recording observations

        Criteria: *Mark*

There are six observations to be made ....1 mark each
                                            Maximum mark 6

Mark Obtained	Skill Level	
	NEA/LEAG/NISEC	SEG/MEG/WJEC
5 and 6	5	3
4	4	2
3	3	
2	2	1
1	1	

Skill: Drawing conclusions

Criteria:	*Mark*

Conclusions:

6. Compares results and suggests that
   rust is formed when air/water/salt is present 3
   ...Needs a little assistance to reach required conclusion............. 2
   ...Arrives at conclusion after much assistance 1
7. Suggests a sensible method of rust prevention from
   the results of the experiment, e.g. greasing 1
8. Suggests any other practical method for
   preventing iron from rusting.. 1

Maximum mark 5

Mark Obtained	Skill Level	
	NEA/LEAG/NISEC	MEG/SEG/WJEC
5	5	3
4	4	2
3	3	2
2	2	1
1	1	1

Laboratory technician's notes

Each pupil involved in this assessment will require the following:

> iron plate (6 pieces, about 1 cm^2 each)
> 6 test tubes in a rack
> safety glasses/goggles
> bunsen burner and mat
> test tube holder
> 4 rubber bungs
> spatula

The following items should be available in the laboratory in a central position:

> boiled/cooled water (about 250 ml)
> oil
> calcium chloride (anhydrous)
> cotton wool
> labels
> grease
> salt water

As this is a fairly complex assignment it would be useful if you remained in the laboratory throughout the lesson. It will be necessary for each pupil's experiment to be retained somewhere safe until the following week. You may need to make temporary arrangements for the storage of groups of test tubes, to release test tube racks for other work. Beakers are often useful for this, or binding groups of test tubes together with elastic bands.

Skills check list – *Following instructions*

Pupil's Name	Step 2 (a) 2	Step 2 (b) 2	Step 2 (c) 2	Step 2 (d) 2	Step 2 (e) 2	Step 2 (f) 2	Total Mark	Skill Level

✳ For each of these criteria, 2 marks should be awarded if the instruction is carried out correctly *without assistance*, and 1 mark *with assistance*.

Skills check list – *Manipulation*

Pupil's Name	Step 2 (a) 2	Step 2 (b) 2	Step 2 (c) 2	Step 2 (d) 2	Step 2 (e) 2	Step 2 (f) 2	Total Mark	Skill Level

✻ **Award 2 marks if an instruction is carried out safely and with good manipulative skill (e.g. no spillage)** *without assistance*, **and 1 mark if carried out safely** *with assistance*.

Assignment 37
Plotting solubility curves

Teacher's notes

This assignment is intended as one to be given to the whole group. It gives a useful comparison of pupils, perhaps at a late stage in the course, and can be used for pupils who have had genuine problems with their assessments (e.g. low skill levels, genuine absences, arriving in the last year at a new school, etc.). It is offered as a written exercise which may be required for purposes of external moderation.

The assignment should be carried out as a follow-up to any teacher-designed solubility experiment. You may wish to change the data offered to incorporate your own findings.

All four graphs need not be assessed, but the best one selected for marking. This means that each pupil is being positively assessed.

Marking scheme

Skill: Presenting data

Criteria:	Mark
Good use of scale on both axes	1 (½ mark each)
Both axes labelled correctly, without assistance	2 (1 mark each)
...with assistance	1 (½ mark each)
Most points placed correctly, without assistance	2
...with assistance	1
Smooth curve attempted/reasonable	1
Maximum mark	6

Mark Obtained	Skill Level	
	NEA/LEAG/NISEC	**SEG/MEG/WJEC**
5 and 6	5	3
4	4	2
3	3	
2	2	1
1	1	

Assignment 38
Investigating the effects of electrolysing purified sea water

Teacher's notes

This assignment could be used during a study of electrolysis. It may prove useful as it tests other areas of the syllabus. It could be expanded to show how the sea contains many valuable minerals and elements which are of use to mankind. It could be linked with social and economic effects of using part of the earth as a resource (e.g. the cost of the energy required to obtain hydrogen, chlorine, sodium chloride, sodium hydroxide, etc.)

One part of the group could carry out the assignment as a normal practical while a small number of pupils have their skills at *Following instructions* and *Manipulation* assessed. These pupils should be grouped close together in the laboratory (see pages 5 and 6).

Hopefully you will be able to provide a real sample of sea water. A plastic lemonade bottleful collected during one Summer Vacation has lasted the author many years, with a little cheating, as necessary! However, it is believed that the important link between resources and the laboratory is established, as suggested above. In fact, it may prove very useful to ask pupils to ask their parents to allow them to collect and bring in small samples from around our shores.

Skills check lists are provided for this assignment on pages 118 and 119.

Marking schemes

Skill: Following instructions

Criteria:	Mark
Sets up filtration	1
Carries out filtration	1
Pours salt water into beaker	1
Connects wires up correctly to voltage supply	1
Places litmus paper over electrodes	1
Puts Universal Indicator into the salt water	1
Maximum mark	6

Mark Obtained	Skill Level		
	NEA/LEAG/NISEC	MEG/SEG	WJEC
5 and 6	5	3	2
4	4	2	2
3	3	2	1
2	2	1	1
1	1	1	1

Skill: Manipulation

Criteria: *Mark*

Folds filter paper correctly.....................................1
Dampens filter paper in filter funnel.....................1
Pours in salt water without spillage......................1
Does not pour over top of filter paper...................1
Connects up electrodes in a safe position...............1
Places litmus paper over electrodes.......................1

Maximum mark..............6

Mark Obtained	Skill Level	
	NEA/LEAG/NISEC	MEG/SEG/WJEC
5 and 6	5	3
4	4	2
3	3	2
2	2	1
1	1	1

Laboratory technician's notes

Pupils will need the following apparatus and materials:

100 ml impure sea water (real, if possible)
filter papers
filter funnel
conical flask
12 volt variable d.c. supply
2 carbon electrodes
litmus paper
Universal indicator
250 ml beaker
two connecting wires
plastic/wood insulating strip to fit across the beaker

This experiment should be ready in the laboratory at the start of the assignment. It should be in place for each pupil being assessed. Other pupils can collect their equipment.

Skills check list – *Following instructions*

Pupil's Name	Sets up filtration	Carries out filtration	Pours salt water into beaker	Connects wire correctly to voltage supply	Places litmus paper over electrodes	Puts Universal indicator into salt water	Total Mark	Skill Level
	1	1	1	1	1	1		

Skills check list – *Manipulation*

Pupil's Name	Folds filter paper correctly	Dampens filter paper in filter funnel	Pours in salt water without spillage	Does not pour over top of filter paper	Connects up electrodes in a safe position	Places litmus paper over electrodes	Total Mark	Skill Level
	1	1	1	1	1	1		

Assignment 39
Investigating the electrolysis of water

Teacher's notes

This is a useful assignment for assessing the whole group. It will enable you to make comparisons between groups of students as well as individual pupils within each group. This may be important in terms of external moderations. The assignment may also prove useful for pupils who have had genuine problems with absence, or the assessments themselves (e.g. a pupil joining the school half way through the course).

Pupils should be made aware that this demonstrated experiment is an assessment and they must act accordingly. At the end of the experiment, disperse pupils around the laboratory to attempt their graphs. Check that they have a reasonable set of results to work from, as the lack of one skill (e.g. taking down recordings) should not be detrimental to another skill being assessed, in this case *Presenting data*.

You will need to explain the important points of this experiment as it progresses. It is important that you do not become lost in the maze of assessment. The prime aspect of this work is still to show what happens when water is electrolysed. For example, pupils still need to be aware as to why sulphuric acid is added to the water. They need to know that water will produce two volumes of hydrogen to one of oxygen, and that this is involved in proving the formula of water.

This experiment may be demonstrated twice during the course. It could be used during a study of water, but it is believed that the assignment is probably best left to a study of electrolysis. Pupils may then find the conclusions easier to formulate.

The 'water' used during this experiment should be approximately 0.5M sulphuric acid, as this provides a quick result.

Marking schemes

Skill: Presenting data

Criteria:	Mark
Labels axes correctly and includes units 2	
Majority of points placed correctly .... 2★	
...some correct 1	
Points are joined up smoothly .... 2★	
...points are joined 1	
Maximum mark 6	

Mark Obtained	Skill Level	
	NEA/LEAG/NISEC	SEG/MEG/WJEC
*5 and 6	5	3
4	4	2
3	3	
2	2	1
1	1	

*These marks must be included to obtain the maximum skill level.

Skill: Drawing conclusions

Criteria: *Mark*

Is able to say:

As voltage increases so does rate of gas given off............................3

...but needs some minimal assistance............................2

...but needs much assistance 1

Correctly identifies both gases 2

Suggest any error which may occur during the experiment..............1

Maximum mark..........................6

Mark Obtained	Skill Level	
	NEA/LEAG/NISEC	SEG/MEG/WJEC
5 and 6	5	3
4	4	2
3	3	
2	2	1
1	1	

Laboratory technician's notes

This is a teacher–demonstrated assignment. It is important that the apparatus is set up for the start of the assessment. The 'water' should be 0.5M sulphuric acid, but should not be poured into the Hoffman's Voltameter. It is strongly suggested that you check the apparatus is working and that the electrodes will be able to stand up to the experiment. The following apparatus is required:

Hoffman's Voltameter set up in a clamp stand in a
    position where it can be seen by a large group
0.5M sulphuric acid (500 ml)
filter funnel
variable voltage supply unit
two leads
platinum electrodes
wood splints
bunsen burner and mat
test tubes and rubber bungs

Assignment 40
Electrolysing copper (II) chloride solution

Teacher's notes

This assignment can be used to assess a small number of pupils for the skill *Following instructions* and *Manipulation*. As this will only involve the setting up of the experiment it is possible for the remainder of the full group to join in after this stage. The rest of the group could be contemplating another written assessment, such as a graph of a design experiment, or they could be doing some research into the present assignment.

It is suggested that this experiment is carried out during a study of electrolysis.

Skills check lists are provided on pages 125 and 126 to assist with the assessment of the two skills.

Marking schemes

Skill: Following instructions

Criteria:	Mark
Connects both electrodes to the power supply	1
Places both electrodes in a beaker	1
Pours in copper(II) chloride solution	1
Places insulation strip between the two electrodes	1
Adds some Universal Indicator	1
Places litmus papers over electrodes	1
Maximum mark	6

Mark Obtained	Skill Level		
	NEA/LEAG/NISEC	SEG/MEG	WJEC
5 and 6	5	3	2
4	4	2	
3	3		1
2	2	1	
1	1		

Skill: Manipulation

Criteria:	Mark
Connects electrodes safely and correctly	2
Pours in solution without spillage	1
Correctly positions insulation strip	1
Dampens and positions litmus papers correctly	1
Maximum mark	5

Mark Obtained	Skill Level	
	NEA/LEAG/NISEC	SEG/MEG/WJEC
5	5	3
4	4	
3	3	2
2	2	1
1	1	

Laboratory technician's notes

The apparatus and equipment required for each pupil attempting the assessment is listed below:

> beaker containing 100 ml copper(II) chloride solution
> 12 volt d.c. power supply
> connecting leads
> a pair of matching copper electrodes
> litmus paper
> Universal indicator
> a suitable insulating strip to separate the copper electrodes

This apparatus and material should be laid out ready for pupils to use. Pupils being assessed should not be expected to collect apparatus during this assignment.

Skills check list – *Following instructions*

Pupil's Name	Connects both electrodes to power supply	Places both electrodes in a beaker	Pours in copper chloride solution	Places insulation strip between electrodes	Adds Universal Indicator	Places litmus papers over electrodes	Total Mark	Skill Level
	1	1	1	1	1	1		

Skills check list – *Manipulation*

Pupil's Name	Connects electrodes safely and correctly	Pours in solution without spillage	Correctly positions insulating strip	Dampens and positions litmus paper correctly	Total Mark	Skill Level
	2	1	1	1		

Assignment 41
Distilling petroleum

Teacher's notes

This has to be a demonstration experiment because of the safety aspects. If carried out under examination conditions, it should be possible to assess all pupils on their ability at *Presenting data* and *Making and recording observations*. The graph required is only a sketch but this can still be a useful assessment.

Each of the fractions collected should be stored until the end of the experiment. Then they can be burnt. The cleanness of the flame should be commented upon.

This assignment would be best carried out during a study of fuels, and be linked to Assignment 42.

No specialised apparatus is needed for this assignment. It may be an occasion when the Liebig condenser could be dusted down.

Marking schemes

Skill: Presenting data

Criteria:	Mark
Labels axes correctly, good use of scale (°C)	2
Plots both sets of points with reasonable accuracy	2
Sketches in graphs, extrapolates	2
Maximum mark	6

Mark Obtained	Skill Level	
	NEA/LEAG/NISEC	SEG/MEG/WJEC
5 and 6	5	3
4	4	2
3	3	2
2	2	1
1	1	1

Skill: Drawing conclusions

Criteria:	Mark
Boiling point increases with molecular size	2
Flame becomes dirtier as fraction increases	2
States two fuels and gives reasons	2
Maximum mark	6

Laboratory technician's notes

This is a demonstration. The teacher will require the following apparatus and materials:

      crude petroleum (about 25 ml)
      a suitable distillation apparatus
      some test tubes with rubber bungs
      a ceramic mat

The apparatus required should be set up in a position in the laboratory where it can be seen by a large number of pupils.

Assignment 42
Investigating the cost, efficiency and cleanness of various fuels

Teacher's notes

This is a design assignment, assessed by written performance, so it can be used as a full group assessment and for the purposes of comparing groups and thus setting internal school standards for moderation purposes.

As with any design assignment, pupils may produce viable alternatives, although they are recommended to use only the apparatus listed. The marking scheme below can be used for guidance, but you should also refer to the general criteria on page 17.

The SEG does not require the assessment of this skill as part of the external assessment procedure. However, it should still be of use in preparing pupils for the formal examination.

Marking scheme

Skill: Designing an experiment

Criteria:	Mark
Suggests using the same weight of each fuel	1
Uses the same amount of measured water each time	1
...as fuel is burnt to heat it	1
Measures temperature rise	1
Mentions need to observe deposits on test tubes	1
Compares amount of heat to cost	1
Maximum mark	6

Mark Obtained	Skill Level	
	NEA/LEAG/NISEC	MEG/WJEC (SEG)*
5 and 6	5	3
4	4	2
3	3	2
2	2	1
1	1	1

*External assessment of this skill is not required by the SEG.

Laboratory technician's notes

If the teacher decides to follow up this assignment with a practical session, you should liaise with the teacher over the equipment and materials needed.

Assignment 43
Distilling coal

Teacher's notes

This assignment can be used to assess only a small number of pupils, preferably no more than eight at one time. The rest of the group can carry on with the assignment as written but not be assessed by it. You could encourage these pupils to assist each other in their development of skills. Those pupils being assessed need to be to one side of the laboratory, close enough to facilitate the assessment, and yet spaced out to ensure that examination conditions prevail.

The assignment requires the use of coal – a rare commodity in many homes today. If you are fortunate enough to find a pupil or colleague with a coal fire, then try to obtain a few large lumps for the laboratory technician to break up.

This assignment could be carried out during a study of fuels, or pollution, or of coal itself as a resource. It should be linked to Assignment 42 if possible.

Skills check lists are provided for this experiment on pages 133 and 134. This may prove to be a useful record in the event of external moderation.

Marking schemes

Skill: Following instructions

Criteria:	Mark
Step 1, adds coal and clamps..............	1
Step 2, places delivery tube................	1
Step 3, places second tube	1
Step 4, places beaker and water	1
Step 5, puts on safety glasses	1
Steps 6 and 7, heats strongly..............	1
Maximum mark	6

Mark Obtained	Skill Level		
	NEA/LEAG/NISEC	SEG/MEG	WJEC
5 and 6	5	3	2
4	4	2	
3	3		1
2	2	1	
1	1		

Skill: Manipulation

Criteria: *Mark*

Correctly places clamp on boiling tube............... 1
Correctly places delivery tube............................ 1
Correctly places collection boiling tube 1
Clamp is manipulated to working position.......... 1
Heats and attempts ignition at same time............ 1
Step 10(e), safely pours and tries to ignite 1
Maximum mark 6

Mark Obtained	Skill Level	
	NEA/LEAG/NISEC	SEG/MEG/WJEC
5 and 6	5	3
4	4	2
3	3	2
2	2	1
1	1	1

Laboratory technician's notes

It is likely that the whole group will carry out this assignment, but only a small number of pupils will actually be assessed. This will probably mean that a large amount of apparatus is required, perhaps as many as fifteen sets, but this is at the teacher's discretion.

Take note of the special delivery tube, which will require constructing if this is not already part of the department's standard experiments (see below). The overflow tube should clear the delivery tube by about 10 cm. This will enable pupils to work safely.

The coal should be broken up into small pieces but not quite powdered. Each experiment will need about half a boiling tube full.

For each pupil attempting the assessment the following apparatus and materials should be in place at the start of the assignment. (Other pupils can collect their own apparatus because they are not being assessed):

clamp stand
2 boiling tubes (which could be kept for future use)
bunsen burner and mat
special delivery tube and overflow (see right)
2 beakers
safety glasses/goggles
wooden splints
Universal indicator
ceramic mat

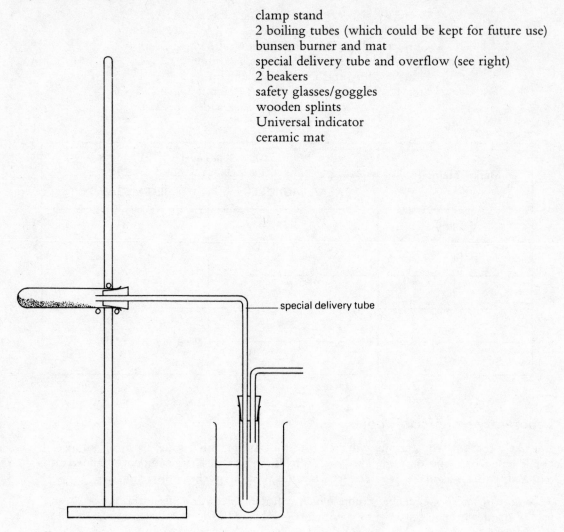

special delivery tube

You may be required to assist the teacher in the laboratory during the course of this experiment.

Skills check list – *Following instructions*

Pupil's Name	Step 1, adds coal and clamps	Step 2, places delivery tube	Step 3, places second tube	Step 4, places beaker and water	Step 5, puts on safety glasses	Steps 6 and 7, heats strongly	Total Mark	Skill Level
	1	1	1	1	1	1		

Skills check list – *Manipulation*

Pupil's Name	Correctly places clamp on boiling tube	Correctly places delivery tube	Correctly places collection boiling tube	Clamp is manipulated to working position	Heats and attempts ignition at same time	Step 10(e), safely pours and tries to ignite	Total Mark	Skill Level
	1	1	1	1	1	1		

Assignment 44
Investigating diffusion of gases

Teacher's notes

You can use this experiment to set a standard, if you wish. As the whole group will attempt the graph, all the pupils can be compared to each other and from group to group.

The experiment could be performed at any time within the overall scheme of work, but it would be best carried out during a study of diffusion. You should also attempt to link the assignment to a study of reaction rate, as it relates to it in terms of how fast particles can move.

The assignment could be extended to include the skill *Drawing conclusions* by asking pupils to look for a relationship between particle size (molecular mass) and the rate of diffusion. It could be useful to extend the graphs to suggest where the two gases could meet if they do not do so during the experiment.

The solutions used should be 2M strength.

Marking scheme

Skill: Presenting data

Criteria:	Mark
Both axes labelled correctly	2
Good use of time scale	1
Most points plotted correctly on both graphs	2
Good shape to graph with points joined	1
Maximum mark	6

Mark Obtained	Skill Level	
	NEA/LEAG/NISEC	SEG/MEG/WJEC
5 and 6	5	3
4	4	2
3	3	2
2	2	1
1	1	1

Laboratory technician's notes

This is a teacher-demonstrated assignment. The following apparatus should be set up in such a place that it can be seen by a large number of pupils:

a diffusion tube 50 cm long, 3 cm diameter
   with two rubber bungs to fit
2M ammonium hydroxide solution
2M hydrochloric acid
cotton wool

tweezers
clamp stand
Universal Indicator papers
metric rule
stop clock

Assignment 45
Designing an experiment

Teacher's notes

This is a design assignment, assessed by written performance, and so it can be done by the whole group under examination conditions. It is suggested that this assignment is set when a study of separation techniques has been made.

It is very difficult to give a marking scheme for a design experiment as variations may occur which are perfectly acceptable. The marking scheme offered below is for guidance only. You may also need to refer to the general guidelines on page 17.

The SEG does not require this skill to be assessed, but you may find the assignment useful when preparing pupils for the formal examination at the end of the course.

Marking scheme

Skill: Designing an experiment

Criteria:	Mark
Logical sequence similar to:	1
Eliminates two jars by taste	1
Shakes up bottles before starting	2
Measures out exactly 100 ml	2
Indicates need to filter	1
Correct method for filtering	1
Retains solid from filter paper	1
Indicates drying solid	1
Weighs solids for comparison	1
Indicates repeating for all bottles	1
Maximum mark	12

Mark Obtained	Skill Level WJEC	
	NEA/LEAG/NISEC	WJEC/MEG (SEG)*
10, 11 and 12	5	3
8 and 9	4	2
6 and 7	3	2
4 and 5	2	1
1, 2 and 3	1	1

*External assessment of this skill is not required by the SEG.

Laboratory technician's notes

You are not required for this assignment.

Assignment 46
An industrial study: air liquefaction

Teacher's notes

This assignment is for general use. It should be set during a study of pollution or industrial chemistry. Although this is a paper exercise and really involves no practical work, it does involve the same kind of processes required by experimental work.

The assignment may prove useful for pupils with genuine absence problems, or who arrive late in the school year. Also, it may be used to moderate different groups within a school, as it can be used as a full group assessment.

Marking scheme

Skill: Drawing conclusions

Criteria:	*Mark*
Question 1	2
2	1
3	1
4	1
5	2
6	2
7	1
Maximum mark	10

Mark Obtained	Skill Level	
	NEA/LEAG/NISEC	SEG/MEG/WJEC
10 and 9	5	3
8 and 7	4	
6 and 5	3	2
4 and 3	2	
2 and 1	1	1

Laboratory technician's notes

You are not needed for this assignment.

Assignment 47
Finding the number of molecules of water in a salt

Teacher's notes

This is a class practical to be carried out by the whole group. Some pupils may be required to carry out the experiment on their own, for the purposes of assessment. It is suggested that a maximum of eight pupils are assessed by this assignment, at any one time.

This assignment could be used during any work being done involving molecular weights and formulae.

You will need to check how hot the crucibles are before pupils attempt to weigh them, for obvious safety reasons.

The assignment has only one skill to be assessed, *Following instructions*. A skills check list is provided on page 140.

Marking scheme

Skill: Following instructions

Criteria:	Mark
Step 1, weighs	1
Step 2, reweighs	1
Step 4, sets up apparatus	1
Step 5, heats/wears safety glasses	1
Step 6, cools/reweighs	1
Step 7, repeats Step 6	1
Maximum mark	6

Mark Obtained	Skill Level		
	NEA/LEAG/NISEC	SEG/MEG	WJEC
5 and 6	5	3	2
4	4	2	
3	3		1
2	2	1	
1	1		

Laboratory technician's notes

The apparatus and materials required for this assessment for each pupil are as follows:

ceramic mat
tripod
fire clay triangle
crucible and lid
a salt (such as copper sulphate)
tongs
balance
safety glasses/goggles
bunsen burner and mat

This apparatus should be in place for those pupils being assessed at the start of the assignment. The bunsen burners should already be burning. Other groups of pupils may require the apparatus; the teacher will advise you about this.

Skills check list – *Following instructions*

Pupil's Name	Step 1, weighs	Step 2, reweighs	Step 4, sets up apparatus	Step 5, heats/wears safety glasses	Step 6, cools/reweighs	Step 7, repeats Step 6	Total Mark	Skill Level
	1	1	1	1	1	1		

Assignment 48
Analysing an unknown chemical

Teacher's notes

This assignment is offered as a 'catch all' assessment. It is a useful assignment in general, but becomes increasingly important towards the end of the course. It can be used for pupils who have missed assignments for genuine reasons. It is a useful way of revising many of the skills and subject areas studied earlier in the course.

Analysis is the one area that brings together many of the skills expected of pupils at this stage. It covers a multitude of skills: *Following instructions*, *Manipulation*, *Making and recording observations* and *Drawing conclusions*.

Following instructions and *Manipulation* can be measured by observing a small group of pupils. The skills *Making and recording observations* and *Drawing conclusions* can be assessed by written response. This means that problem pupils can be assessed for four skills at once.

It is envisaged that the whole class will do the assignment. This may mean sharing apparatus, but as long as examination conditions prevail then it is a viable assessment.

It would be useful if you demonstrated an analysis before the assignment and suggested to pupils that it does not matter where they start the assignment. All the tests are to be done and negative responses must be made (e.g. this is *not* a sulphate). Pupils must name the unknown chemical as the final conclusion.

Marking schemes

Skill: Following instructions

Criteria: *Mark*

There are six practical instructions
(Steps 2 to 7) to be carried out................1 mark each
                        Maximum mark................6

Mark Obtained	Skill Level		
	NEA/LEAG/NISEC	SEG/MEG	WJEC
5 and 6	5	3	2
4	4	2	
3	3		1
2	2	1	
1	1		

Skill: Manipulation

Criteria: Mark

There are six mini-stage experiments from
Steps 2 to 6..1 mark each
Maximum mark................6

Mark Obtained	Skill Level	
	NEA/LEAG/NISEC	MEG/SEG/WJEC
5 and 6	5	3
4	4	2
3	3	2
2	2	1
1	1	1

Skill: Making and recording observations

Criteria: Mark

There are seven observations to be
recorded. Mark any six correct ....1 mark each
Maximum mark................6

Mark Obtained	Skill Level	
	NEA/LEAG/NISEC	SEG/MEG/WJEC
5 and 6	5	3
4	4	2
3	3	2
2	2	1
1	1	1

Skill: Drawing conclusions

Criteria: Mark

There are 7 mini-conclusions to be
made in each Step ....1 mark each
Names unknown chemical correctly,
...with no assistance 3
...with minimal assistance 2
...with much assistance 1
Maximum mark...............10

Mark Obtained	Skill Level	
	NEA/LEAG/NISEC	SEG/MEG/WJEC
9 and 10	5	3
7 and 8	4	2
5 and 6	3	
3 and 4	2	1
1 and 2	1	

Laboratory technician's notes

Each pupil being assessed will require the following items set up before the assignment starts:

> bunsen burner and mat
> safety glasses/goggles
> spatula
> flame test rod
> test tube holder
> test tubes in a rack
> distilled water
> sodium hydroxide solution (1M)
> hydrochloric acid (1M)
> barium chloride solution
> silver nitrate solution

If the whole class is to undertake this assignment it might be necessary to share equipment.

Your presence in the laboratory would be helpful for this assignment.

NOTES